New Top Show Flat
顶级样板房[IV]

本书编委会·编

中国林业出版社
China Forestry Publishing House

图书在版编目（CIP）数据

顶级样板房.4/《顶级样板房》编委会编. — 北京：中国林业出版社，2013.4

ISBN 978-7-5038-6993-8

Ⅰ.①顶… Ⅱ.①顶… Ⅲ.①住宅－室内装饰设计－作品集－中国 Ⅳ.①TU241

中国版本图书馆CIP数据核字(2013)第055080号

编委会成员名单

策　　划：思联文化

编写成员：贾　刚　刘增强　高囡囡　王　超　刘　杰　孙　宇　李一茹
　　　　　姜　琳　赵天一　李成伟　王琳琳　王为伟　李金斤　王明明
　　　　　石　芳　王　博　徐　健　齐　碧　阮秋艳　王　野　刘　洋
　　　　　陈圆圆　陈科深　吴宜泽　沈洪丹　韩秀夫　牟婷婷　朱　博
　　　　　宁　爽　刘　帅　宋晓威　陈书争　高晓欣　包玲利　郭海娇
　　　　　张　雷　张文媛　陆　露　何海珍　刘　婕　夏　雪　王　娟
　　　　　黄　丽　程艳平　高丽媚　汪三红　肖　聪　张雨来　韩培培

采　　编：柳素荣

责任编辑　纪　亮
文字编辑　李丝丝

出　版：中国林业出版社（100009 北京西城区德内大街刘海胡同7号）
网　址：http://lycb.forestry.gov.cn/
E-mail: cfphz@public.bta.net.cn
电　话：（010）8322 5283
发　行：中国林业出版社
印　刷：北京利丰雅高长城印刷有限公司
版　次：2013年5月第1版
印　次：2013年5月第1次
开　本：230mm *300mm　1/16
印　张：19
字　数：200千字
定　价：320.00元（USD 60.00）

样板生活　生活样板

室内设计是实用艺术，样板房设计的实用性就尤为明显。一直以来，样板房及销售中心都是销售的道具，成熟的地产公司对自己所推的楼盘的样板房都有非常完整的规划，很多时候，设计师只是用自己的专业知识配合地产公司将他们需要营造给客户的家居梦造出来。在这样的一个系统中，设计的重要性其实并没有那么明显，重要的是设计之前的调查、分析、定位等前设计的工作。这并不是故意贬低设计师的重要性，地产公司还是非常注重明星设计师的效应，经常不惜重金聘请明星设计师如梁志天等为其设计样板房。非常有意思的一个现象是这些明星设计师设计的样板房往往并不如一些不那么知名的设计师所设计的样板房好卖。地产商找这些知名设计师的目的行非常强，他们要的就是他们那些成功的符号，而不知名的设计师比较了解当地的居住习惯、消费习惯，设计的样板房更受消费者追捧就是非常自然的事情了。这些都反映了样板房设计的实用性的诉求。这是一个梦想与真实交织的地方，设计师为普通的消费者营造了仿佛触手可及的家居梦，让普通的消费者坠入其间，迅速决定长期持有这一梦想家居。这样的样板房，从商业的角度而言，无疑是成功的，但是我们更希望看到那些实在一些，再实在一些的样板房设计，在这些样板房设计中，我们可以看到户型如何优化等更加务实的设计内容，而不是天花乱坠的造梦。

"顶级样板房"是一套系列丛书，该系列已经持续出版了4册。本册收集了近年来活跃在当下的一线设计师的最新作品，这些作品既有著名设计师的经典再现之作，也有不那么知名设计师的务实之作，希望读者们可以从中既学习到造梦的技法、说故事的能力，同时也不忘思考务实的问题，唯有如此才可以设计出不但视觉上惊艳而且住起来也舒服的家来。这才是我们真正向往的样板生活。

编者
2013年4月

Contents

The Liquid Space ★ 液态空间 006	Function ★ 机能至上 054	The Crown of the East ★ 东方之冠 106
Curve ★ 曲线 012	Private Salon ★ 专属奢华 060	Wang Jinyuan ★ 望今缘 114
Harmony and Balance ★ 和谐与平衡 016	Exquisite Life ★ 精致生活 068	Fashion Brand Shop ★ 时尚名店风 122
Residence 8 – Leaf ★ 8号院——树叶 022	Ideal Home ★ 都会理想居 076	Modern Simple Style ★ 梦想照进现实 130
Residence 8 – flower ★ 8号院——花 030	The Cage ★ 笼子 084	Modern Neoclassical Style ★ 混搭的视觉盛宴 136
Cheif Cook ★ 行政总厨 036	Curios ★ 多宝格 090	Warm Sunshine ★ 光之暖 142
Slowing Down the Pace of Life ★ 私享慢生活 042	Cut and Fit ★ 剪切与合适 094	British Style ★ 古典英伦范 148
Encounting Arts ★ 都会艺遇 046	Clean and White ★ 极简·白 100	Crush ★ 碰撞 156

目录

Dynamic and Pure ★ 动感与纯粹的融合 . . . 162	Colors ★ 彩条复兴 . . . 210	Classic Noble ★ 古典尊崇 . . . 256
Simple and Oriental Style ★ 简洁东方 . . . 168	Courtyard Garden ★ 天井花园 . . . 216	Low-Key Luxury ★ 低调的奢华 . . . 264
Modern City Fashion ★ 都会时尚 . . . 172	Elegant Space ★ 素雅格调 . . . 224	Private Garden ★ 私密花园 . . . 270
Northern Europe Style ★ 北欧情怀 . . . 180	Zen ★ 禅意东方 . . . 228	Clouds ★ 云海之居 . . . 276
New Oriental ★ 新东方主义 . . . 184	Water Flows, Wind Breezes ★ 水殿风来 . . . 234	Xi'an Family ★ 熙岸世家 . . . 284
Modern City ★ 新巴洛克主义 . . . 190	Simple Chinese Style ★ 简约中式风 . . . 240	French Style ★ 法式风情 . . . 294
Culture and Life ★ 文化栖居 . . . 198	Butterflies in Gray Shadow ★ 灰影蝶舞 . . . 244	The Post 80S ★ 80后主张 . . . 300
Nature's Charm ★ 自然旖旎 . . . 206	American Modern ★ 美式摩登 . . . 250	

The Liquid Space
液态空间

According to the original design, the indoor space will be shown before our eyes in the form of "water". Water exists in nature in liquid form. The purpose of the design is to bring us a flowing feeling, which will be felt through the floorboards, walls and even the hang ceilings. An extending effect is created and the whole space seems to be covered by liquid.

Oak is used in order to better show the liquid design, completely connecting the design on the second floor to the first one. The design style spreads from floorboards of the first floor to the walls and the hang ceilings, then it continues from the stair well to the floorboards, the walls and the hang ceilings of the second floor. Curve-style furniture and lighting decorations are purposely selected to integrate with the design theme.

项目名称：北竿山国际艺术中心B-1别墅样板房
设计单位：KLID达观国际建筑设计事务所
设 计 师：凌子达、杨家瑀
项目地点：上海
建筑面积：280 m²
主要材料：橡木染灰、米白色珍珠漆、爵士白大理石、碳色不锈钢
摄 影 师：施凯

最初的设计概念是以"水"的形式去表现室内空间。水在自然界中是以"液态"的形式存在着,它会流动在地板,墙面,甚至在吊顶上,形成延伸的感觉,好像是液体包覆了整体空间。

我们选择了橡木来表现这种液态造型,并使这液态造型从二楼到一楼整个都连起来,从一楼的地面延伸到了墙面、吊顶,再从楼梯间延伸上去,到二楼的地面、墙面、吊顶。家具和灯具饰品也选用了弧线造型的样式,更加融合这设计的主题。

Curve 曲线

Curve, mostly used in the design, can be reflected through the cambered wall and the light from the hang celling. It is also used in the facade to create stereo effect. In the dinning hall, a U-shaped space is created with the proper design of chairs and their backs. In this way, not only a unique shape is formed, also the space of the dinning hall is defined. The stair beside the dinning hall looks like a sculpture, of which the handrail looks like a perfect curve wriggling from the third down to ground floor.

The application of materials is simple, and white is used as the main color. The combination of both simple materials and colors perfects the curve.

项目名称：北竿山国际艺术中心B-2别墅样板房
设计单位：KLID达观国际建筑设计事务所
设 计 师：凌子达、杨家瑀
项目地点：上海
建筑面积：350 m²
主要材料：爵士白、白色珍珠样、碳色不锈钢
摄 影 师：施凯

整体设计概念是以"曲线"作为设计的一个主轴。"曲线"的设计应用在弧形的墙体上和吊顶的光带中,在立面上也用了曲线形成了立体的层次,在餐厅的部分也把座椅和靠背形成了一个大的U型空间,不仅是做了一个造型(Form),也界定出餐厅的空间。餐厅旁的楼梯好像一个雕塑,它的扶手是一条完美的曲线,从三楼一直蜿蜒到一楼。

为了清晰地表现出曲线的造型,材料的应用十分单纯,是以"白色"为主,纯净的材料与色彩应用,才能表现完美的曲线。

Harmony and Balance
和谐与平衡

A nature palette is for some people who love to live with fresh open air, clear sky, the smell of the rising sun, purity, harmony. That inspiration is from the widespread and deep-rooted desire of city dwellers to bring nature into the urban environment.

By accepting and responding to the given architectural nature of the space is key to the Architectural palette. The critical decisions involve deciding how to alter the way natural light enters the space; developing a lighting strategy; and placing objects in the spaces.

The colors and materials of the nature palette are drawn from a gentle spectrum of natural hues-pale, sun and sea-bleached wood, and sand and pebble grey on floor finish. Combined rose metal is a great way to create an illusion of depth and space.

Lighting is essential to creating a calm, airy mood, and unifying large spaces so they can be read as completed volumes. Besides, the architectural lighting will reinforce the shape of the architecture, the walls, the lines in the building as well.

项目名称：华润置地南通橡树湾样板房
设计单位：刘伟婷设计师有限公司
设 计 师：刘伟婷
项目地点：江苏南通
建筑面积：180m²
摄 影 师：冯志毅

顶级样板房 | Top showflats

· 平面布置图

户外新鲜的空气,晴朗的天空,太阳的味道,纯洁以及和谐,这是大自然给人们的天然享受。把这种乡村美景带到城市生活的环境中,这种灵感来自于城市居民的普遍愿望且根深蒂固。

接受并考虑环境的建筑性质,是建筑装饰材料自动调色系统的关键。关键在于如何改变自然光进入空间的方式,以及如何设计照明方式,如何放置房间里面的物体等。

自然调色板的颜色和材料来自于自然柔和色温范围的光谱,太阳、海漂木、沙子、卵石灰色地板刷油与玫瑰色合金一起,创造深度和空间上的幻觉。

对于轻松和愉快的心情,以及将整个空间一览无余,照明设备是不可缺少的。另外,建筑采光凸显建筑风格,墙面的轮廓线条感十足。

Residence 8 – Leaf
8号院——树叶

As a theme of "leaf", designer uses leaf pattern as a partition to divide the living area and resting area. The panel gives privacy to the owner but not blocking the whole area. Spacious feeling can be kept. Also, during the movement of the sliding panels, the graphic is changing alone.

The most dominant material is wood in the unit but there are many different kinds of woods to make contrast. To prevent the unit being too warm, some mirror and metal are added as a highlight.

设计单位： Ptang Studio Ltd.
设 计 师： Phillip Tang
项目地点： 辽宁 大连
建筑面积： 120m²
主要材料： 不锈钢、云石、墙纸、布艺
摄 影 师： Ulso

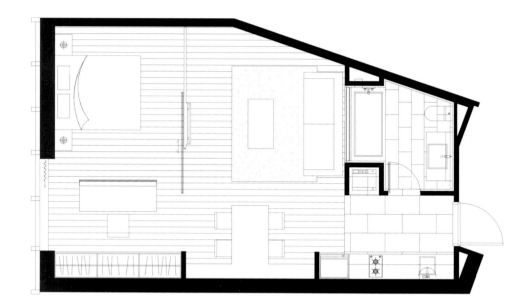

·平面布置图

"树叶"成为设计的主题。设计者通过运用树叶状的隔断将客厅和休息区分开。这种形式的隔断既保证了主人的隐私，又不至于封闭了整个区域，让你有一种看上去显得挺宽敞的感觉。此外，在移动滑动隔断时，你感觉到的只是隔断面板上面的图画发生了变化。

这个单元使用的主要材料为木材，通过运用不同种类的木材形成对比效应。为避免房间过于暖和，我们在设计时加入了一些镜子和金属，这也是本单元的一个亮点。

Residence 8 – Flower
8号院——花

Flower is the main theme of the design. Various methods, including the use of veneer, leather and baldric, are used to set off the beauty of flowers by designers.

The large semipermeable wooden screen partition wall between the bedroom and the living room will be the most attraction in the design. It is magnified in design style. When opened, it can not only act as a partition, also it will make you feel that it is not the connecting of space, but the expansion. The overall color is natural and soft, with the decoration of fabric sofa and soft baldric, giving you a strong feeling of a warm home. The four hanging pictures perfect the design theme, creating a fashionable Chinese style.

The characterized bed head looks like a falling and flowing flower, giving stronger visual effects. It not only riches the overall design atmosphere, also highlights the architectural art.

设计单位： Ptang Studio Ltd.
设 计 师： Phillip Tang
项目地点： 辽宁 大连
建筑面积： 120m²
主要材料： 不锈钢、云石、墙纸、布艺
摄 影 师： Ulso

顶级样板房 | Top showflats

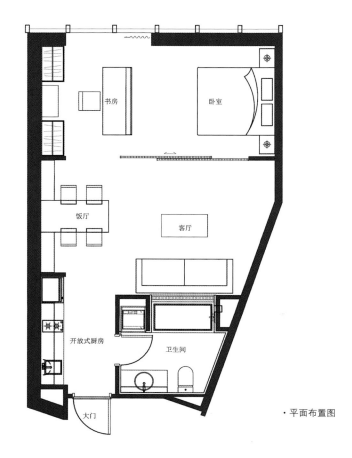

· 平面布置图

整个设计以花为重点,设计师利用木皮、皮和配饰等,配合用不同的手法来表现花的形态,突出整体主题。

睡房和客厅的大型半透木屏风间隔,顿时变成焦点,设计上刻意夸大,这样不但可以将两边区分,敞开时更不会感到这是空间的过度而是空间的扩张。整体色调自然柔和,配合布艺沙发部分软装配饰,营造一个温暖的家的感觉!四幅挂画起了点题作用,形成一个中式的时尚格调。睡房的特色床头有如花瓣飘落,形态流动,大大增强视觉效果,既丰富了整体气氛又可突出艺术感!

Cheif Cook
行政总厨

To give the first impression into the house, there are various kinds of animals shaped furniture on the first floor and demonstrating the owner's interest obviously. In the living room, there are stainless steel "chopping boards" stacking to form the coffee table, "plates" piling up to form the table and "cow" shaped area rug; while in the dining area, there are "noodles-liked" pendent lamp, the whole first floor is interesting.

Furthermore, in the bar area, forks were beat to form a side chair and there are round "sheep-liked" bar stools. The pendent lamps are made up of milk bottles, the study table is in "marshmallows" shaped and the wash basins are inspired by the shape of the "egg cup" as well. In the master bedroom, the wardrobe is similar to the old style refrigerator with the big size metal hinges and pumping hands and the dressing table is using the idea of the food warmer using on buffet table. In the master bedroom, designers make use of forks and plastering, and turn these forks into interesting feature wall.

项目名称：晶苑四季御庭F2户型样板房
设计单位：壹正企划有限公司
设 计 师：罗灵杰、龙慧祺
项目地点：上海
建筑面积：280m²
摄 影 师：罗灵杰

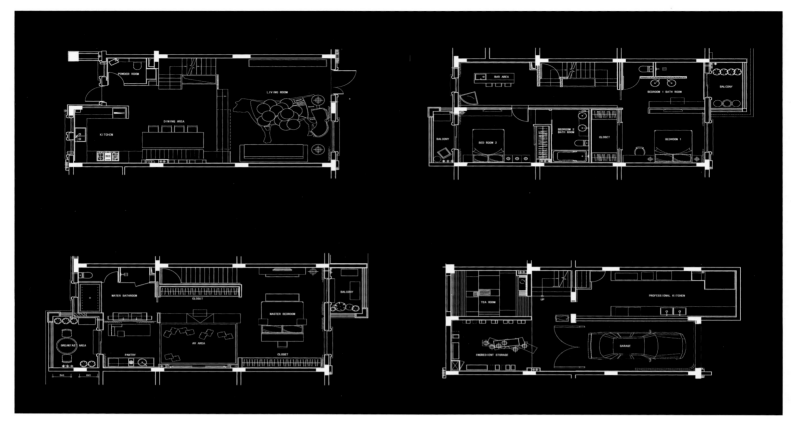

· 平面布置图

从进入客厅开始，已经看到一些以"猪"为形态的矮台、用镜钢造成的一块块"砧板"一样的茶几、以餐碟堆砌而成的角几，以及地毡亦有如牛型状的设计。及后来到饭厅，用仿铜面条造成的灯具，令整个首层充满了玩味及特色，成功让人能融入这个用烹饪及美食所营造的气氛。

二层水吧区，用了叉子掘成椅子，加上"吧椅"为圆型，感觉似一只只绵羊，在抽象中体现主体；在睡房吊灯用上奶樽型的玻璃樽造成，书台型状似棉花糖；而洗手间里也精妙地以"蒸蛋器"造型设计成洗手盆。在主人房间，衣柜造型与旧式雪柜相似，配有大型金属铰及抽手。墙身的特色墙用上一些叉子，幻化成飞镖般随意粘贴于墙面，营造形式别致的画面及带出不一样的惊喜玩味效果。另外，模仿旧式雪柜及自助餐炉的形态所造成的衣柜与梳妆台，亦突破了惯常家具的形态框框，创意中更见高贵格调，十分独特。

Slowing Down the Pace of Life
私享慢生活

Regarding the apartment layout, this case has two bedrooms and one living room. The spatial function is divided according to the demands of young people and that the designer rethinks the new living model of space and living style. The design has many elements of life and exhibits a new oriental layout to us. The designer dismantled the wall adjacent to the dining room and kitchen so that the kitchen forms an open area while the cabinet surface extends and amazingly turns into a table. The small reading space nestles in the corner of the living room, allowing you to enjoy the slow life. The reading room creates a flexible transition for private space in the conversion process, and offers more possibility for family interaction. The balcony is specially designed for a dining area that when combined with the outdoor landscape, make one space.

In the material and color, the designer expects to extend the line to the outdoor landscape so that the color stays cleaner. Combined with the color of black, white, and grey, the soft decoration and outdoor landscape bring a fresh feeling. Thus, we can take off our inhibitions without being self-conscious and taste the dialogue of life and the environment.

项目名称：成都中德英伦联邦项目
设计单位：广州市柏舍装饰设计有限公司
项目地点：四川成都
建筑面积：100m²

顶级样板房 | Top showflats

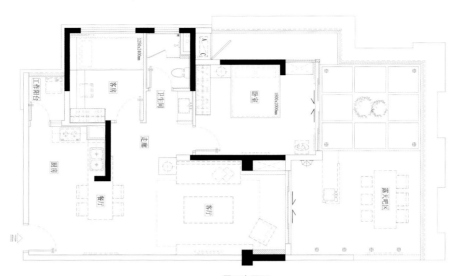

· 平面布置图

在平面布局优化上，本案为两室一厅的户型，空间功能以年轻一族的社交需求来划分，是设计师通过重新思索空间与生活方式连结的新时代住宅形式。设计师把生活元素融入到规划中，重新发展出一种属于东方的住宅平面：布局上拆除了餐厅与厨房固有的相邻墙体，让厨房区域得到释放，而橱柜的台面延伸，顺理成章地成为了餐桌。这种既开放又独立的多重空间关系，创造出一种空间与生活的对话状态，赋予生活一种视觉的层次。而隐于客厅一角的阅读空间，面积不大，却让人向往着闲时千金难买的私享慢生活，为开放空间到私密空间的转换创造一个具有弹性的过渡枢纽，同时也为家庭的互动提供了另一种可能。露台特别设置的户外用餐区，巧妙地借用了户外的园林景致，使空间整合为一。

在材质与色彩的运用方面，设计师期望将居住者的视线延伸至户外景观，于是特意让全屋色彩保持纯净，并运用黑白灰的经典配搭，结合室内软装，还有室外郁郁葱葱的生态环境蕴生出的清新感觉，内外呼应和谐。身处其中，不自觉地褪去了繁嚣都市的拘束感，品味生活与环境的对话。

Encounting Arts
都会艺遇

It lies in the most prosperous area, the heart of Zhongshan District, Taipei. The future residence enjoys an easy access to the downtown of the city and closely neighbors the business center. All you need in your daily life can be satisfied in the neighboring area and it is quite convenient in transportation. It is the best living place, which highlights the value and character of the residence. .

The space style of the sample house is based on metropolis art and you can imagine it as both a warm house and a private hostel for successful persons. It makes innovations in the partition wall and makes the most use of the space, perfecting the indoor space in an open and transparent way. Materials used in the wall of living rooms are steel baking veneer. Materials used in edges of walls are stainless steel titanize skirting board and smallpox modeling, extending the space. Stepping into the room, you can fully understand the delicate processing of details and feel the extension and layers of the space. The presentation of artworks deepen the design and artistic connotation of the sample house.

项目名称：甲山林台北一号苑
设计单位：动象国际室内装修有限公司
设 计 师：谭精忠
参与设计：黎咏嘉、陈敏媛
项目地点：中国台湾台北市
建筑面积：248 m²

顶级样板房 | Top showflats

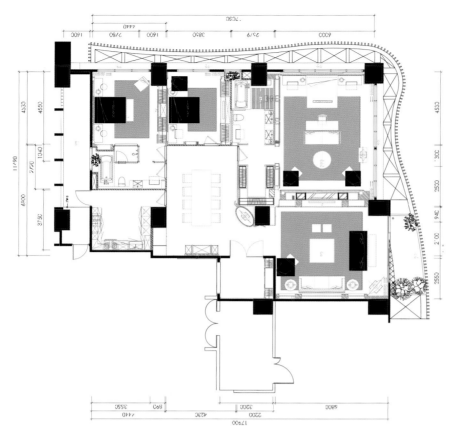

· 平面布置图

本案位处于台北市中山区正中心精华地段，未来建筑基地坐拥市中心地段，周边各项生活机能完整，交通网路便捷，紧邻商务中心，是理想的居住环境，更突显本案的价值及独特性。

样板间空间风格以都会艺遇作为主轴，成功人士的住家融合私人招待所的概念为空间发想。打破传统隔间墙的作法，重新思考空间的可能性和极大值，以开放、穿透的手法，将室内空间极致化。公共区壁面材质使用钢烤木皮涂装，不锈钢镀钛踢脚板与天花造型收边，带出空间的延续性。进入室内便能了解细节的精致处理并感受空间的张力与层次感。艺术品的呈现更加深了本案的设计深度及艺术涵养。

Function
机能至上

The interior design of an apartment shows the occupant's desire and ambition for his life. This design for the apartment highlights the intimation and reconciliation between space and human beings and aims to bring the functions of each unit into full play. It does not have any unnecessary frills but sets out the low profile integrated design and subtle details. Its kinetonema exhibits comfort and facility of each function in the space. It adopts the materials and arrangement with visual penetration to extend various unit spaces.

Entering the living room through vestibule, the extensity is amplified. The designer, with his consistent style, endows the space with latitude. The style of dining room keeps identical with that of living room, laying a dark color round table with the matching chairs. Both sides of the hallway seem solid, but the secret mini kitchen drawer is hidden behind.

The laminating cloth on the main wall of master bedroom is simple but elegant matching the tactile sensation which is fit with its function of sleeping. The high cube container, in possession of both foundations of show case and container, contains cleverly a dressing table, which keeps the space integrated. The semi-transparent yarn clamping glass used on the video wall and the ironware hiding the indirect lighting make the space more stretching.

项目名称：新润都峰苑B2户型样品屋
设计单位：动象国际室内装修有限公司
设 计 师：谭精忠
参与设计：庄舜杰、何芸妮
项目地点：中国台湾台北市
建筑面积：208 m²

顶级样板房 | Top showflats

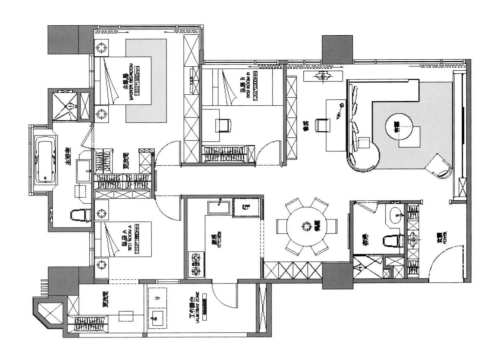

• 平面布置图

一个居家型态的构成，表现出居住者对自我生活的期盼与想像，本案设计重点强调空间与人之间的紧密依存，主要特色是以机能作为出发点，没有过度的装饰，强调不夸耀的整体设计与繁复细节的重现；动线上强调生活机能之流畅感，运用视觉穿透建材及语汇，延展多样化单元空间。

通过玄关，进入客餐厅，空间感整个放大。餐厅延续客厅的风格，使用深色圆形餐桌与同系列餐椅。端景之双侧看似封闭式造型，实为厨房mini暗门柜。

主卧室强调材质质感之睡眠区，床头主墙裱布，材质素雅。而因应空间尺度需求规划出整体造型，展示、收纳兼具化妆功能的高柜，巧妙将化妆桌功能隐藏于高柜中，保持空间完整性；电视墙借由半穿透夹纱玻璃，运用铁件搭配隐藏间接灯光，使空间更显延伸。

Private Salon
专属奢华

This quarter is located at the center of Xin Zhuang District of Xin Bei City. It is a new plan at the nice area with convenient communications and mature living functions. It has a great environment for urban life. The overall space is designed by low profile and refined classical style, matching exclusive materials with containing character and sedate and elegant colors for expanding space dimension. It is inspired by the notion of private hostel and builds the atmosphere for banquet, embellished by artworks to create a distinguished and elegant residence filled with art and humanity.

The design for the master bedroom presents comfort and poise. The wallhanging and leather are bottomed on the main walls, the totem on which is divided by the ironware at the key points to emit the mild light. It makes the master bedroom elegant and distinctive. The usage of brushing veneer paired up with the leather and gray glass on the clothing containing cabinet makes it more delicate. The semi-transparent texture of the yarn clamping glass panel matching the mild light create a kind of special visual enjoyment differed from normal wardrobe. The utilitarian containing function in the wardrobe reflects the elegant life style.

项目名称：新润都峰苑样品屋
设计单位：动象国际室内装修有限公司
设 计 师：谭精忠
参与设计：赖欣慧、林青蓉
项目地点：中国台湾新北市
建筑面积：307 m²

· 平面布置图

本案位处于新北市新庄区正中心精华地段,为拥有顶级地段、便利交通及完善生活机能的新建个案,是都会生活中理想的居住环境。整体空间运用低调而洗练的古典语汇,搭配内敛质感的材质与沉稳优雅的色调,型塑空间大器感。并以私人招待所的概念为出发点,营造出迎宾宴客的情境氛围,辅以艺术品点缀空间的独特性,缔造富有艺术人文的雍雅居所。

主卧室的设计用舒适且大器的基调来呈现,在重点墙面皆以壁布及皮革为基底,并在重点处运用铁件镭切割图腾,内透柔和的灯光,呈现出主卧室的雅致与独特。另外,在衣物收纳柜方面,运用钢刷木皮并且搭配皮革与灰境,不同材质的运用提升了精致度;而夹纱玻璃门片则呈现半透明质感并辅以柔和的灯光,营造别于一般衣柜的视觉韵味。柜内实用贴心的收纳机能,反映出优雅的生活模式。

Exquisite Life
精致生活

This case is located at the cross of the Second Phase of Linkou Culture Road and Zhongxiao Road, the new indicator of Linkou center. It is a new quarter with convenient communications and mature living functions. The sample house adopts the style of low profile and refinement with poised and exclusive materials and colours assortment to create its expanding dimension and show the key of its design intent: a stage coexisted harmoniously with delicate life, furniture and art.

The open configuration in the living room and dining room makes the space ample and lofty. The wide vision with stretch of 10 metres shows the majestic temperament of top villas. The cinereous marble paved floor in public space, in sharp contrast to the white spray painted wall, reflects its poise. The regular separation delimiter and dark colored brushing veneer demarcate the space by visual impression and highlight the space tension.

All cabinets in the living room and dining room are hidden in the wall, which follow the impression of separation delimiter and make the space clean and clear, and, furthermore, is a good accommodation area. The paintings hanging on the simple spray painted white wall reflect the elegant atmosphere which strengthen the impression that the house full with art.

项目名称：林口世界首席样品屋
设计单位：动象国际室内装修有限公司
设 计 师：谭精忠
参与设计：陈姿蓉、许玉臻
项目地点：中国台湾新北市
建筑面积：172 m²

顶级样板房 | Top showflats

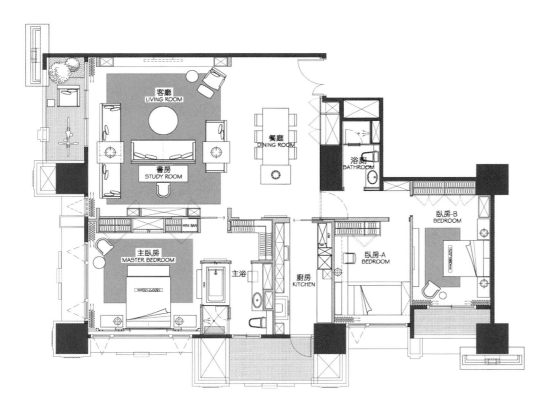

• 平面布置图

本案位处与林口文化二路及忠孝路口，为林口新都新指标，便利交通、完善生活机能的新建个案。样品屋以低调而洗练的设计语汇，内敛而富质感的材质与色调搭配，型塑空间大器感。

由入口进入客、餐厅区时，入内即能感受贯穿客厅与餐厅之开放式空间的宽敞与气度，面宽近10m宽开阔的空间视野，呈现大器豪宅气势；公共空间地坪铺设灰黑色之大理石石材，与白色喷漆壁面形成强烈的对比，进而衬托出空间的沉稳与内敛。墙面规律性的分割与染深色的钢刷木皮，不仅将空间区域性以软性手法加以分界，更加以铺陈空间的张力。

客、餐厅区域之柜体均使用暗柜方式处理，延续墙面的分割效果同时，隐身在墙面后方的柜体，不仅使整体空间线条干净，更有高机能的收纳空间。素白的喷漆墙面吊饰画作艺术品，衬托出空间里优雅的氛围与品味，强化了住宅与艺术结合后的加分效果。

Ideal Home
都会理想居

This case is located at the golden mile in the Zhongshan District of Bei City, which is one of the rare quarters containing fine area, convenient communications and mature living utilities. It has a great environment for urban life. Its style of low profile and refinement with poised and exclusive materials and colours assortment creating an expanding dimension show the key of its design intent: a stage coexisted harmoniously with delicate life, furniture and art.

The overall wall in the vestibule is combined with the surface integrated by linear boards. The built-in cabinet which can accommodate both clothes and shoes is hidden in the simple decorating wall. It hides the original structure and is very utilitarian.

As long as the person enters the area of living room and dining room through the vestibule, its amplitude and lofty quality of the open space penetrating the living room and dining alcove can be felt. The innovative express of design passing through the living room, dining room and kitchen and the new orient style assorted with linear board, wainscot and frame express the power of top villas. The cave stone matched with simple shaped built-in stone fireplace is used on the main wall of the living room. The wall with brushing veneer ornamented by the linear board and stone skirt has some exclusive visual enjoyment with delicate details processing. The space tension and stereovision are laid out.

项目名称：尊胜（白金苑）样品屋
设计单位：动象国际室内装修有限公司
设 计 师：谭精忠
参与设计：詹惠兰、陈敏媛
项目地点：中国台湾台北市
建筑面积：347 m²

顶级样板房 | Top showflats

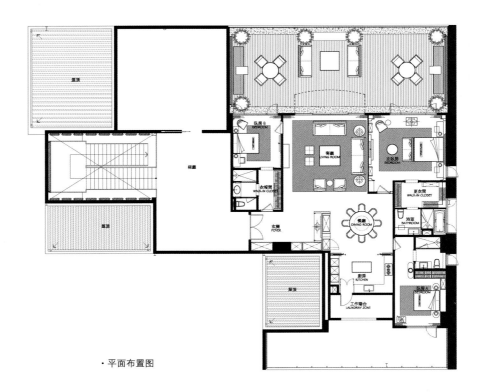

· 平面布置图

本案位处台北市中山区精华地段，为少数拥有顶级地段、便利交通及完善生活机能的新建个案，是都会生活中理想的居住环境。本案的设计重点：精致生活、家私艺术、极致共容的 展演舞台。

玄关以整体壁面与线板整合为面的连结性，展开进入样板房的序曲。而在简练的造型壁板内，另藏有兼具衣帽与鞋子的壁橱收纳功能，也将原有的结构巧妙地隐藏在壁橱中，对应收纳空间的高度需求外，更具备了绝对的实用性。

由玄关进入客、餐厅区时，入内即能感受贯穿客厅与餐厅之开放式空间的宽敞与气度，以新思维的设计表现手法来连贯客、餐厅及厨房，搭配运用线板、壁板、斗框等演绎新东方语汇，呈现大器豪宅气势；接待客厅主墙以洞石搭配线条简练的崁入型石材壁炉，钢刷木皮质感的壁面造型在线板与石材踢脚的点缀下，创造出独有的视觉韵味，也突显出精致的细节处理，铺陈空间的张力与层次感。

The Cage
笼子

Unite I - design professional couple with a young girl

The designer unit focuses around a central, double height space above the dining room, where the space is defined by a wooden screened "cage". This gives the overall architecture a unifying theme, making this high ceiling space more intimate. The material offered as interior finishes are warm but chic at the same time. They offer a beautiful backdrop for the ultra-modern furniture spread across the different rooms in the unit. Many of the furniture are iconic modern pieces, defining the modern artistic and design taste shared by the couple who lives there. Within the living spaces one sees traces of the owner's design profession becoming part of their home decoration.

项目名称：万科第五园样板房一单元
设计单位：如恩设计研究室/Neri&Hu Design and Research Office
设 计 师：郭锡恩、胡如珊
项目地点：上海
建筑面积：350m²
主要材料：橡木贴面、黑色花岗岩、熏黑的青铜、钢丝网、彩色玻璃、marmorino、墙纸
摄 影 师：Zhonghai Shen

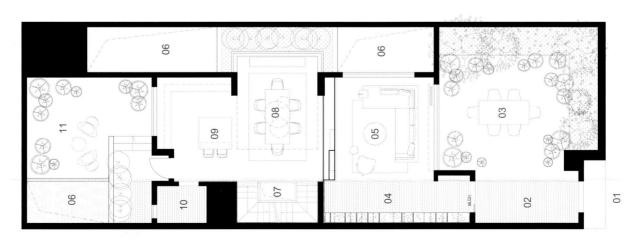

· 平面布置图

第一单元——三口之家:设计师夫妇和女儿

设计师将餐厅上方极具魅力的贯穿两层的空间作为视觉中心。一个木质屏风的"笼子"将它从整个居住空间中分隔出来——这给整体室内建筑定义了一个统一的主题,同时也使二层的空间更具私密性。这一单元的内部空间色调呈暖色系,同时又非常时尚雅致。设计师为精心摆放在不同房间的超现代家具设计了美丽的背景。这些家具中很多都是抽象的现代装饰,居住在这里的设计师夫妇与大家一起分享他们的现代艺术和设计品位。这一空间主人的设计品味和风格已成为了家居装饰的一部分。

Curios
多宝格

Unit II – Three-generation family living in a contemporary house

Derived out of the "DuoBaoGe" concept from Chinese furniture, this house unit utilizes its "side strip", defined as a storage shelving unit and circulation, as the characteristic element of this house. The three generations are each given a storage unit, and each exhibit different hobbies/obsessions on their designated shelves. On the ground level by the living room, the grand parents showcase their collected porcelains and pottery. On the basement level next to the child's play and learning area, the child displays his artwork and paintings. On the third level master bedroom, the parents collect their calligraphy tools and other artifacts.

The remaining spaces are decorated in a more conservative, tailored scheme of furniture and collectives, showing a very livable ensemble of designed furniture and lighting.

项目名称：万科第五园样板房二单元
设计单位：如恩设计研究室/Neri&Hu Design and Research Office
设 计 师：郭锡恩、胡如珊
项目地点：上海
建筑面积：390m²
主要材料：橡木贴面、黑色花岗岩、熏黑的青铜、钢丝网、彩色玻璃、marmorino、墙纸
摄 影 师：Zhonghai Shen

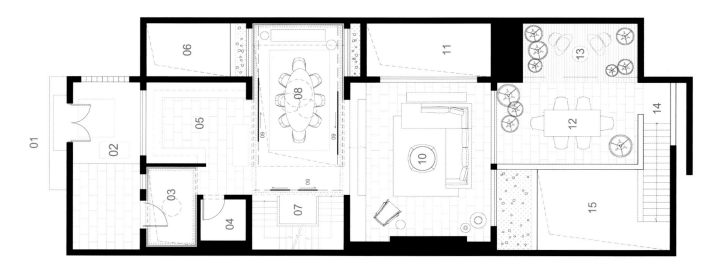

· 平面布置图

第二单元——三代同堂

灵感来源于中国的传统家具"多宝阁",这个居住空间的每个小单元都运用了"方形地带"的概念,奠定了空间结构的特色——如同规制的储物空间。居住在这里的三代人都有他们各自的储物空间,他们的储物架上都有各自为之着迷的收藏品。在一层的客厅里可以看到祖父母收藏的瓷器和陶器;地下一层紧挨着学习和玩耍区域的,是孩子们艺术和绘画作品的展示;三层的主卧里有父母收藏的书法等其他艺术作品。

这一单元其余的空间则由更为传统、量身定制的家具及收藏品系列进行装饰,呈现出非常舒适的居住环境和照明的整体效果。

Cut and Fit
剪切与合适

We think every home must be noble and aristocratic in volume. This doesn't mean it has to be huge, even the smallest rooms become superior when they are based on good proportions. We custom this space by "Cut and Fit" Simple lines and forms with lighter hues for today's home.

Key design features, such as a series of rose shiny partitions, intricate bronze fins and a series of bespoke shimmering architectural light adorn the venue to provoke a sense of movement and motion. Set against an understated backdrop of pale timber and clam wallpapers, the visual experience is further enhanced with bespoke furnishings in accents of costly cacao, champagne and Windsor wine.

项目名称：华润置地泰州国际
设计单位：刘伟婷设计师有限公司
设 计 师：刘伟婷
项目地点：江苏 泰州
建筑面积：120m²
摄 影 师：冯志毅

顶级样板房 | Top showflats

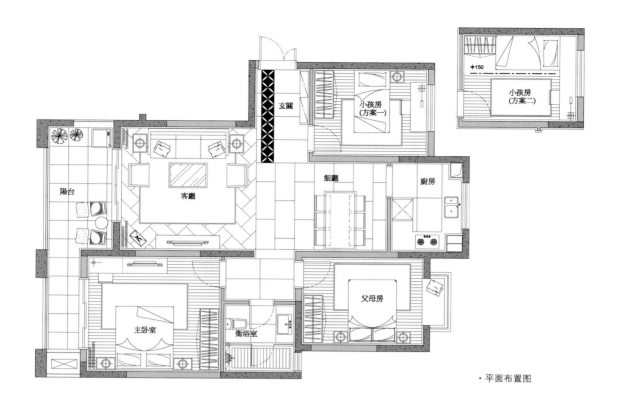

· 平面布置图

我们从每一个家庭都是高尚的贵族去构思。这并不意味着它必须拥有巨大的空间，即使是小小的房间，只要有良好的比例，也可具备成为高尚空间的特色。

我们自定义空间"剪切和合适"，简单的线条亮丽而成今日的家居。关键的设计特点，如一系列的玫瑰金作闪亮的分区，富内涵的铜制物料加闪闪发光的建筑灯系列。在木材和壁纸的背景衬托下，加强眼睛的视觉体验。我们利用光的反射，美丽的泉源充满创造惊喜。

Clean and White
极简·白

The design of the model houses is originally for a two-floored compound apartment with two bedrooms, a living room and a dining room and what has been allocated to our party is to design a layout with four bedrooms, a living-room, a dining room and two bathrooms. In the plane layout, we could see that all rooms are with large-area lighting windows and the spatial arrangement is designed based on the premise of comfort and large space. The adoption of the mobile furniture increases the flexibility of the space. No partition is designed to separate the living room and the dining room in the first floor, which absolutely increases the area of the first-floor public areas. While the second –floor public area of activities connects each private space and it acts as a buffer zone to enter into the second floor by stairway from the first floor. The connection of the study room with the public area will contribute to a multifunctional space after the combination of both areas.

This project is characterized with white space based on minimalism and it calls for a spacious and bright space. Therefore, no much special modeling is designed on the walls and ceilings, while it adopts the architectural method to separate the large space. The main materials adopted are white emulsion paint, white painted plates, white wood floor and white-oak finish. The material is required with integrity and completeness so as to express clearly and concisely the structural relationship of large surface. Some of the wall surface is required to apply special material to firm wall textures, for example: design the background of the parlor sofa with the adoption of white stones in various sizes, which are arranged to form a concise but not simple gradient wall texture; While it adopts some white cotton and hemp ropes in different thickness as the background of the master bedroom to have a horizontal layout and this will reflect the integrity without lacking the details and won't bring an artificial impression.

项目名称： 世欧澜山样板房
设计单位： 广州市东仓装饰设计有限公司
设 计 师： 梁永钊
参与设计： 黎颖欣
项目地点： 福建福州
建筑面积： 130m²
竣工时间： 2012.12

顶级样板房 | Top showflats

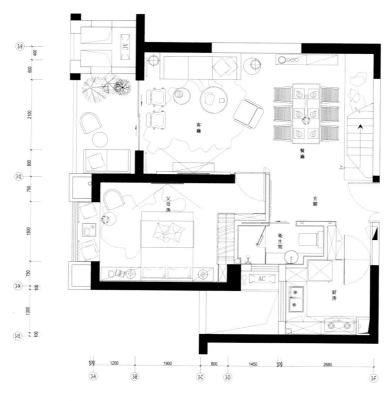

· 一层平面布置图

本案以现代西方极简主义为主题,结合新时代生活模式,颠覆以往中高端住宅停留在消费者心中的奢华形象。本案面向中高端消费群体,三四十岁的成功人士家庭为主流,主要是五口之家,在这年龄层段他们崇尚的是简洁舒适的居家模式,注重家人间的沟通,代表着新一代的生活观念及审美观念。此方案提出现今中高端住宅所欠缺的极简主义,吻合主流消费群体的需求,少即是多,带来一种耳目一新的设计概念及居住模式。

样板房设计项目原为两房两厅的两层复式住宅,我方设计为四房两厅两卫配置。平面中全部房间均有大面积窗户采光,空间布置以舒适宽敞为大前提,运用活动家具增加空间灵活性。一层客厅与餐厅无隔断分隔,使一层公共区域更为宽阔。二层公共活动区域衔接了各个私密空间,并且是一层楼梯进入二层的一个缓冲区域。书房与公共活动区域衔接,组合可成为一个多功能空间。

本案是极简主义的白色空间,空间要求宽阔明亮,因此墙身天花并无太多特殊造型,运用建筑的手法,进行大空间的分割。主要材质采用白色乳胶漆、白色烤漆板、白色木地板及白橡木饰面。材质讲究整体完整,简洁明了地表达大块面的结构关系。部分墙面运用特殊材质塑造墙面肌理,如客厅沙发背景运用的是大小不一的白色石头,构成一幅简洁但不简单的墙面渐变肌理,主人房背景运用的是粗细各异的白色棉麻绳横向排列,整体但又不缺乏细节,并无矫揉做作之感。

顶级样板房 | Top showflats

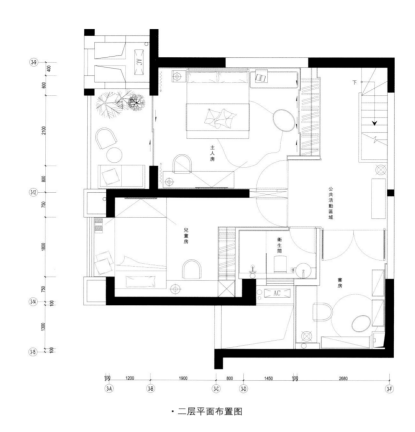

• 二层平面布置图

The Crown of the East
东方之冠

Living room: The ceiling with the design style shaping like the Chinese character "hui", which symbols money, and the thickened large peony wall with three-dimension designed, form a focus of vision. It highlights the easterners' pursuit of perfection. The pattern planning of the house is not only based on the space center to demonstrate its extraordinary power, also it combines more with the noumenal advantage of the building. The room space extends from the non- barrier dinning hall, the study, corridor and outdoor scenery, creating a powerful presence.

Dinning room: The noble and elegant marble stone column defines the space for living room and dinning room. The round-shaped ceiling and the arrangement of the furniture create a relaxed dining atmosphere, representing the meaning of reunion and perfection.

Master bedroom: The focus of the design of the master bedroom with complex space is the arc major wall with unique shape. The tiered tactile sensation not only enriches the layers, but also it extends outwards, liking the warmth brought by embracement of hands. The simple and snow-white master bedroom changing room demonstrates its white and brightly changeable features with the thick guy cloth board and furniture of delicate lines.

项目名称：台中东方之冠样品屋
设计单位：天坊室内计划有限公司
设 计 师：张清平
项目地点：中国台湾 台中
建筑面积：595m²
摄 影 师：刘俊杰、赖寿山

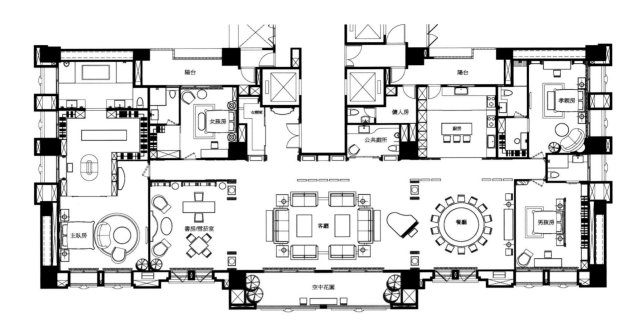

· 平面布置图

客厅中象征钱币的回字型天花及加厚立体设计的大幅牡丹墙，在宽阔的空间中形成视觉凝聚，点出东方人文的极致风华。格局规划除了立于空间中心点，呈现非凡的大器度外，更结合建筑本体的优势，以无阻隔的餐厅、书房、回廊、户外景观为延伸，成就出运筹帷幄的磅礴气度。

用高贵优雅的大理石柱区分定义客厅与餐厅空间，天花采用圆型设计加上家具的陈设，营造轻松的用餐气氛并传达出团圆、圆满之意。

多进式的主卧以造型独特的弧型主墙为设计焦点，层叠质感让层次感丰富，更以延伸而出、如手般的环抱带出拥抱般的温馨。简洁纯白的主卧更衣室，以丰厚绷布板及线条精致的家具，发挥出白色的璀璨多变，让房间充满了雍容华贵的复古氛围，显出高人一等的气度与经典百老汇般的时尚意味。主浴方面空间十分宽敞，设计风格以南洋spa为主调，巧妙地以透明玻璃及马赛克，让设计感不言自明。

Wang Jinyuan
望今缘

Top villa has been so far the tiptop of its kind. The sample room of Wang Jinyuan stands for the highest architectural and design level of a country or region with its style of " Top complanation villa". It also reflects the ideal living style of elite class in our society.

The existence of this kind of top villa is important and necessary. There were so many top villas in domestic and foreign history. For example: the large castles of nobles and celebrities in western countries, the quadrangle courtyard of nobles in ancient Beijing, the grand family courtyards of rich merchants in Shanxi and Anhui. In modern times, top villas in China are rare, of which the most famous are Violet Garden and Sandalwood Place in Shanghai . However, there were lots of top villas in western countries in modern times, which are extremely unique in design styles. These villas are mainly customized and designers combine the surroundings with requirements of house owners. Compared with these villas, Wang Jingyuan not only owns more covered areas and courtyard areas, also it owns more functions and more luxurious decorations both inside and outside of the room. It Seeks for extreme comfort and luxury, and shows its extraordinary design characteristic. Besides, the detail processing represents the quintessence of "Top complanation villa" with different design ideas and manifestations.

项目名称：成都郎基今缘样板房
设计单位：天坊室内计划有限公司
设 计 师：张清平
项目地点：四川 成都
建筑面积：490m²

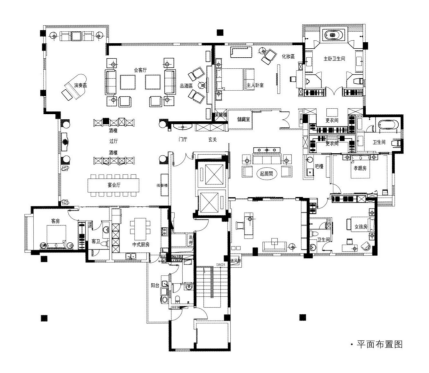

· 平面布置图

豪宅是目前全球居住建筑的最高端类型,望今缘的样板房以"顶级平面化别墅"的格局,来代表了一个国家或地区居住建筑开发、设计的最高水准,同时也反映出社会菁英阶层的理想生活方式,创造其存在的重要性和必然性。

历史上国内外豪宅都屡见不鲜,如欧美等国名流显贵占地广阔的庄园城堡,北京王公贵族庭院重重的四合院,山西、安徽等地富商巨贾气势磅礴的家族大院等。当代国内豪宅就凤毛鳞角了,其中上海紫园和上海檀宫较有影响。而当代欧美发达国家却有相当数量的极富特色的豪宅出现。这些豪宅多是由设计师根据不同的环境特点和业主需求而量身定制的精品住宅。与这些豪宅相比,望今缘使建筑和庭院面积更大、功能更复杂、内外装更豪华。追求极端的舒适、豪华,并展现与众不同设计特色。而且在许多细节处理方面以不同的设计理念和表现形式展现"顶级平面别墅"的精粹。

Fashion Brand Shop
时尚名店风

Zhong Hai Shijia is both fashionable and elegant. The noble of the city is demonstrated by designers' application of the fashionable shop style. Designers follow the fashionable shop styles to reflect the beauty of the city, which means it has integrated with the world tiptops.

For those who value lifestyle, concise and grand rooms are always their prime choice. Based on this, we lay emphasis on the consistency between style and color. High style can be sensed from the elegant neutral color. The simple line, mainly used to decorate the space, simplify all the complicated design. Exquisiteness lies in simplicity. The designers are very careful during the processing of each decorative article. Those decorations without being exaggerated exactly show the essence of fashion and create comfortable and warm atmosphere.

项目名称：中海世家
设计单位：德坚设计
设 计 师：陈德坚
项目地点：江苏 苏州

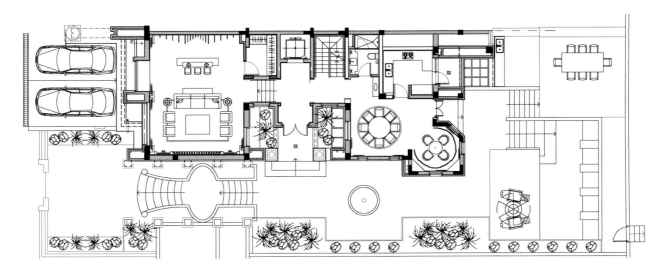

· 一层平面布置图

中海世家的设计时尚优雅,设计师以时尚名店风来表现城市的贵气,再融入品味来表现城市现代感,意味着与世界尖端接轨。

对于重视生活品味的人士,简洁中具有气派一向都是他们首选,故此风格及色调上强调一致性,中色色调的优雅感觉下却不失高格调的气息。以简单的线条作为主要的架构来描写空间,将一切繁复的设计简单化。简单里呈现细腻,设计师在处理每一件修饰的时候非常细致用心。就是那没有过分刻意的修饰,正好能展现时尚的品味,也营造了舒适良好的气氛。

Modern Simple Style
梦想照进现实

Modern simple style would build indoor space which adapt to modern life by utilizing new material and technology, seeking for main features both of simple and straightforward such as valuing utilization ration of indoor space, emphasizing on close cooperation between function division of indoor space and arrangements of the furniture, proposing to abandon additional decoration both of redundant and fussy as well as following fashion on color and modeling.

Areas of the 2 bedrooms is not large, so designers would design the units to be a space of design philosophy which looks like simple while not losing both capacious and comfortable. To move wall of the kitchen on the entrance to inside place, so that porch ark could be embedded masterly nearby the flue, which would not only make function division of whole space be reasonable but also give consideration to functions of use, and bring a kind of home enjoyment to customers.

This prototype room is defined to be of a modern style, plane material of decoration such as irregular geometric solid, white baking finish, mirror surfaces, stainless steel had been utilized for the whole. The floor on living room was paved with stone, which worked concert with the background wall of geometry. There are marble, mirror surfaces

项目名称：沈阳长春华润橡树湾样板间
设计单位：阿珞室内设计（北京）有限公司
设 计 师：张颂光
项目地点：吉林长春南部新城
建筑面积：71m²
主要材料：石材、爵士白、灰木纹、太阳冰花木饰面、白色混油钢琴漆、墙纸、不锈钢、马赛克
摄 影 师：史云峰

and beautiful colored oil painting on wall space of the kitchen, matching with classical chair and round table, which will make intense comparison to both echoes with the color and forms. The background wall of soft roll at the bedroom continued to use the style of wall of geometry at the living room, adding changes, which had made it be linear. The wardrobe of transparent glasses had taken in both dress and personal adornment of master in a glance, which would further highlight both style and grade of master. Warm color sheets, nostalgia pattern cushion and carpets, night table of classical color zebra-stripe pattern both of black and white, antique photo frame, which had added some tender feelings of home in light green tonal background. At children's room, combination furniture scatters randomly, which will not only satisfy children's requirements both of study and rest but also show both hobbies and interests of children.

Whole prototype both mixed and matched furniture containing retro structure through briefness arrangement and stuff of crisp tone, the distinctive is the tone of furniture will give one very strong visual impact, combining with retro accessories again, which will perfectly show appreciating pictures both of mixing and matching.

现代简约风格通过运用新材料、新技术来营造适应现代生活的室内空间,力求以简洁明快为主要特点,重视室内空间的使用率,强调室内的功能分区与家具布置之间密切配合,主张废弃多余的、繁琐的附加装饰,在色彩和造型上追随流行时尚。

此二居室面积不大,所以设计师将此户型设计为一个布局看似简单却不失宽敞舒适的设计理念空间。将入口处厨房的墙壁内移,这样在烟道旁能够巧妙地嵌入玄关柜,既使整个空间功能分区合理,同时又兼顾了使用功能,给客户带来一种家的享受。

此样板间定义为现代风格,整体运用不规则几何形体、白色烤漆、镜面、不锈钢等装饰面材。客厅地面满铺的石材地面与几何形背景墙形成呼应。餐厅用大理石、镜面和色彩艳丽的油画做背景墙面,配以古典椅子和圆桌,形成强烈的色彩呼应和形式对比。卧室的软包背景墙延续了客厅的几何背景墙风格并加以变化,线条感十足。透明的玻璃衣柜使主人的服饰一览无余,更加凸显主人的格调和品位。暖色的床单、怀旧图案的靠垫和地毯、经典黑白两色斑马纹图案的床头柜、古董相框,在淡绿色调的背景里平添了一丝家的温情。儿童房内高低错落的固定家具组合既满足了孩子学习与休息的需求,同时也展示出孩子的兴趣爱好。

整个样板间通过布置简洁、色调清爽的硬装,混搭着含有复古构建的家具,与众不同的是家具的色调给人很强的视觉冲击,再次与复古的配饰品相结合,完全演绎了一场混搭的视觉盛宴。

Modern Neoclassical Style
混搭的视觉盛宴

Design style of modern neoclassical is an improved classical style, more seemingly like a way of thinking of diversification; using modern skill and material quality to restore classical temperament, which will combine romantic themes which meditate on the past and modern requirements for life, which is compatible with luxury and elegance as well as fashion and simplicity, from simple to miscellaneous, from whole to parts, fine craft will give one a meticulous impression.

The prototype room is arrangements of classical 2 bedrooms in recent years, so there are not excessive alterations in space of prototype room, there are only more reasonable divisions on utility function of owners, for example, design porch ark in the only space of hallway; open style design both of the kitchen and dining room; design to make space be more transparent through wardrobe of glasses at master bedroom; children's room not only had fields both of living and study but also had functions such as playing and exercises in limited space.

There are diamond modern elements which are fully utilized on decoration to combine with neoclassical texture material. Modern skills of irregular diamond on walls both at drawing room and dining room continued to be used

项目名称：沈阳长春华润橡树湾样板间
设计单位：阿珞室内设计（北京）有限公司
设 计 师：张颂光
项目地点：辽宁沈阳市于洪区
建筑面积：78m²
主要材料：石材、灰木纹、蓝木纹、雕刻白、木饰面、秋香木、白色混油钢琴漆、壁纸、不锈钢
摄 影 师：史云峰

顶级样板房 | Top showflats

· 平面布置图

on background wall at master bedroom and walls nearby the desk at children's room to highlight features both of modern and fashion, both stainless steel bars on TV background wall and unevenness design of blue wood grain stone had highlighted features which are both simple and miscellaneous of modern neoclassical as well as from whole to parts, which would interpret combinations both of neoclassical and modern elements.

The premise of giving considerations to luxury and elegance as well as fashion and simplicity, the color of stuff was defined as neutral color, so ash wood grain stone was selected and used, blue wood grain stone matching with autumn incense wood, and adding embellishment of hard material such as stainless steel, mirrors at Qing dynasty, tawny glasses, which will mix together with neutral color to the point. Integral furniture and decorations had interpreted neoclassical temperament through gray tone both of black and white, single chair of the table was designed to highlight gray tone both of black and white by using both front and back of cloth, the embellishment of throw pillow in sofa had made the whole be more vivid and tactile sensation as well as more vitality, rushes of Bedding at master bedroom had set off an honorable sensation both of noble and elegant, spotlight is in wardrobe which is designed by using both stainless steel and glasses, there are pearl jewelry, which are treasured by a hostess on the platform of python skin in the wardrobe, leather product inverts its images in pictures of mirrors, which will set off master noble grade through python skin of classical temperament. Human engineering design is reasonable utilized in the wardrobe at children's room, putting toy district at lowest end, which will work concert with playing fields. Blue lint, white lacquer as well as smart and vivid color of beddings had made children's room be clean, lively. Baseball accessories with sports elements had stimulated both hobbies and cultivations of children, making children enjoy the happiness which was brought by baseball star A-rod with bathing under the blue sky.

现代新古典主义的设计风格是经过改良的古典主义风格，更像是一种多元化的思考方式；用现代的手法和材质还原古典气质，将怀古的浪漫情怀与现代人对生活的需求相结合，兼容华贵典雅与时尚简约，从简单到繁杂、从整体到局部，精雕细琢都给人一丝不苟的印象。

此样板间为近年来经典两居室布置，故样板间并未在空间上做过多的改动，仅从业主的实用功能上做了更合理的划分，比如在门厅仅有的空间里设计玄关柜；厨房和餐厅的开敞式设计；主卧玻璃衣柜使空间更通透的设计；儿童房在局限的空间内既具备了生活学习区域，又兼备了儿童玩耍活动功能等。

在装饰上充分地运用了菱形的现代元素，与新古典质感的材料相结合。客餐厅墙面不规则菱形的现代手法，沿用到主卧背景墙与儿童房书桌台墙面，突显了现代时尚的特点，电视背景墙面不锈钢条，与蓝木纹石材凹凸造型的设计，突显了现代新古典的简单与繁杂、整体到细节的特点，完美地诠释了新古典与现代元素的结合。

兼顾华贵典雅与时尚简约的前提，硬装颜色定义为中性色，故选用了灰木纹石材、蓝木纹石材和秋香木的搭配，并加以硬朗的材料，如不锈钢、清镜、茶镜的点缀，与中性色相融合的恰到好处。整体的家具与饰品均以黑白灰色系来诠释新古典的气质，餐桌单椅设计师巧妙地运用布料的正反面来突出黑白色灰色系，沙发中抱枕的点缀使得整体更加生动与质感，更具有生活气息，主卧床品的棕色毛片衬托给主人带来一种高贵典雅的尊贵感，射灯照在不锈钢与玻璃设计的衣柜中，柜中蟒蛇皮的台面上放着女主人心爱珍珠饰品、皮具倒影在镜子的画面中，与古典气质的蟒蛇皮衬托中主人的高贵品位。儿童房的衣柜合理的利用人体工程学设计，将儿童玩具区放在最下端，更与玩耍区域相呼应。蓝色绒布与白色漆木以及床品的俏皮鲜艳的色彩让儿童房显得干净、活泼，具有运动元素的棒球饰品能激发孩子对运动的爱好，让孩子沐浴在蓝色的天空下享受着与棒球明星A-rod一起运动带来的快乐。

Warm Sunshine
光之暖

The three-storey foreign-style house, simply designed, is matched with unique artworks. It is proper in design proportion. The orderly layout of the rooms will not block your visual line. A spacious and comfortable home comes to your eyes. The full height glass windows are specially reserved to let in the outdoor scenery and sunlight. Different feeling and atmosphere will come to the house based on the different light during the daytime and night.

The skillful use of different kinds of materials makes a skillful layout. Wooden floors are used in many parts of the room, warming the house. A black mirror is placed on the wall which spreads from the second-floor corridor to the minor hall. Images reflected in the mirror is loom, creating a mysterious atmosphere and widening the space in the foreign-style house. Besides, colorless glass is placed in stair position to enrich the visual effect. And the fairly high decorative bookcase between the study and living room highlights the unique height of the space.

项目名称：比华利山示范单位
设计单位：Danny Cheng Interiors Ltd.
设 计 师：郑炳坤
项目地点：中国 香港
建筑面积：380 m²
摄 影 师：Danny Cheng Interiors Ltd.

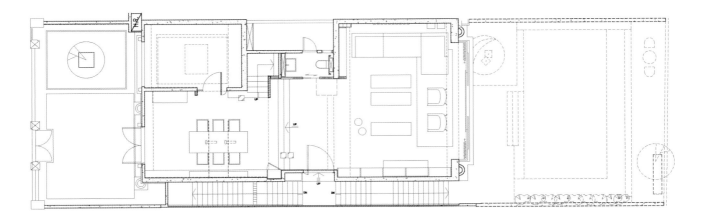

· 一层平面布置图

这个共三层高的洋房,简约的设计配以别致的装饰品,而且设计上比例合宜,井然有序的布局让视线贯穿整个单位,务求做一个宽敞且舒适的家。特意保留其落地玻璃窗,窗外美景尽收眼底,亦让室外充足的光线带进屋内,而且随着日与夜不同的光线,为屋子带来不一样的感觉和气氛。

巧妙的物料运用,利用不同的物料营造出一个巧妙的布局。全屋多个位置均铺上木地板,增添温暖感。于二楼走廊伸延至偏厅的墙身装上黑镜,镜上反映出来的景象若隐若现,加添了一点神秘感,而且令洋房的空间感进一步提升。 另外,楼梯位置亦装上清玻璃,令视觉画面更为丰富,还有客厅及阅读间的超高装饰书柜,更突显着室内其非凡独有的楼高。

British Style
古典英伦范

  Let's return to the origin of life; modern design means are adopted to interpret the classic British style. Based on the original and traditional British special layout, the artistic colors of blue, grey and green endow the room with the sense of rhythm and beauty. The high-ceiled lobby and comfortable dinning rooms seem more elegant together with the comfortable and large American furniture and hand-made accessories. The red-dominated underground audio-visual room is set off by the velvet texture. Billiards activity area surrounds the place, bringing about better interaction between the entire space and the owne.

The master bedroom features the sense of layer. As the private space of the owner, it is marked by great functions and comfort. In terms of soft ornament, it is decorated by unified colors, warm and soft fabrics. The changing room built in the master room perfectly interprets the luxurious atmosphere. You don't have to go to Paris or Milan since the resplendent stage is right here.

The designer turns the space into a life-oriented one. This dream-brewing attic is transformed into the multifunctional soul harbor of the female owner who is a costume designer, so that she can gives into full play her own inspiration in such a free space. After a busy day, she can get herself relaxed in the SPA, cosmetics and YOGA areas.

项目名称：淮南领袖山南样板房E户
设计单位：大勻国际设计中心
设 计 师：陈亨寰
参与设计：陈雯婧
软装设计：上海太舍馆贸易有限公司
客　　户：安徽省淮南市天恒置业有限公司
项目地点：安徽
建筑面积：700m²
撰　　文：刘慧瑛

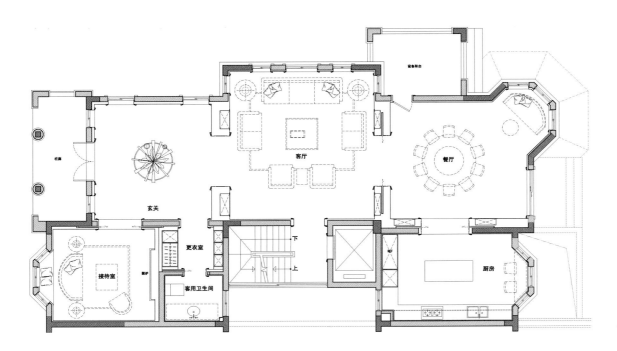

· 一层平面布置图

顶级样板房 | Top showflats

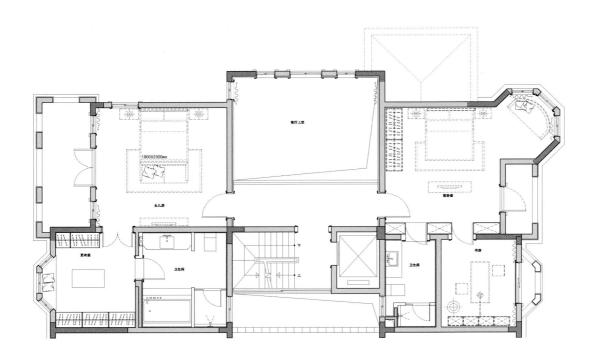

· 二层平面布置图

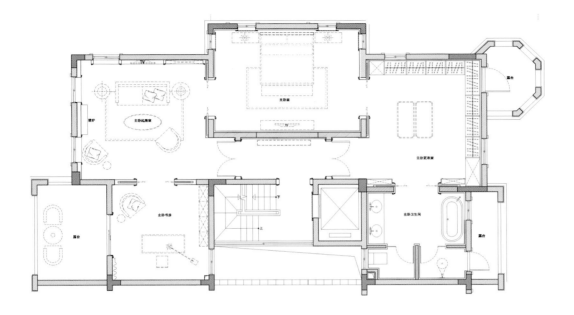

· 三层平面布置图

回归生活的最初点,本案试图用现代的设计手法阐释古典英伦,在原有传统英式住宅空间格局下,以蓝、灰、绿富有艺术的配色处理赋予室内动态的韵律和美感,挑空的大堂及舒适的餐厅配以舒适的大尺度美式家具及手工质感的小饰品,更显品位。以红色为主色调的地下视听室,采用丝绒质地将整个空间烘托得更妖娆多姿。周边设有台球活动区域,让整个空间与主人更好地互动起来。

主卧室强调空间的层次与段落,作为主人的私密空间,主要以功能性和实用舒适为主导,软装搭配上用色统一,以温馨柔软的布艺来装点。主卧配套的更衣室,将奢华大气演绎到极致。不在巴黎,也不需前往米兰,此处即是最华贵的秀场。

设计师将空间赋予更多的生活化,我们将这个造梦的阁楼,设定成女主人的多种用途的心灵空间。女主人作为服装设计师,在这一亩天地中,尽情发挥自我灵感,在忙碌过后,内设的SPA、美容、瑜伽区域,更可带来放松心灵的无尽体验。

Crush
碰撞

Fashionable and classical style have been used in this sample house. The style with rich layers and strong tactile sensation not only divide the space, but creates the artistic effect with strong visual effect. The design theme of the sample house is fashionable, edgy and characterized. Black and white, mainly used in the room, is matched with dark blue. Some parts of it are dotted with mental decorations, which go with the leather materials, luxurious silk materials and mental tawny glass. It is simple, grand and delicate.

The villas are three-storey buildings. The living room, dinning room and old-people room are in the ground floor. The living room, dinning room and open-type western kitchen are in a same space. The theme color is peaceful blue. The western large fireplace, the three-dimensional carving bookshelf on the wall, and cupboard decorations, are fully coated with dark blue. The fashionable furniture, matched with leather sofa, features the host. The modern and concise dining table and chairs look vigorous in the calm space.

The second floor functions as master bedroom, boy's room and guestroom. The wall of the lift hall is covered with large decorative pictures.

项目名称：成都万科五龙山3C户型样板房
设计单位：北京空间易想艺术设计有限公司
项目地点：万科五龙山公园内
建筑面积：460m²

顶级样板房 | Top showflats

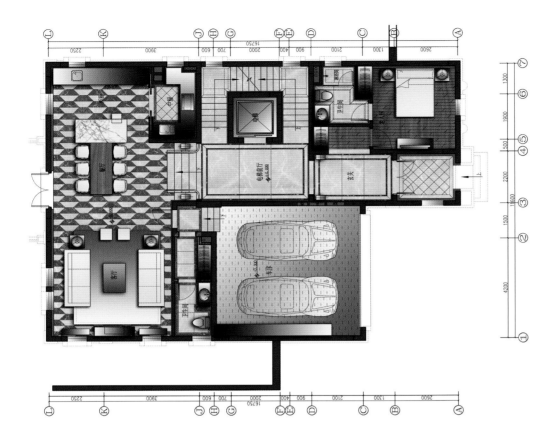

· 平面布置图

本案运用了现代与古典的碰撞，层次丰富并且质感强烈的造型，这些造型不仅为空间划分了区域，同时又努力创建富有视觉震撼的艺术效果。时尚、前卫、个性成为本案的设计主旋律。整体色调整体以黑、白色调为主，配以深蓝色，局部点缀金属材质饰品，搭配皮草面料和奢华的丝绒面料，配合金属质感的茶几，简约大气又不失精致。

本案别墅建筑为三层结构。一层主要为客厅、餐厅及老人房。客厅和餐厅及敞开式西厨整体为一个空间，主题采用宁静的蓝色为基调。欧式的大型壁炉、墙面的立体雕刻书架以及橱柜等装饰都大面积采用深蓝色主题。摩登时代的家具配以皮质沙发，彰显主人个性。餐厅区现代、简约的餐桌餐椅在沉稳的空间中突显灵动。

二层主要功能为主卧室、男孩房及客房。电梯厅墙面采用了巨型装饰画铺满墙面，在现代气息中点缀少许古典，突出空间气势与文化内涵。

Dynamic and Pure
动感与纯粹的融合

The whole space is in modern and modest style. With the application of space division method, the whole space is divided into several proportional parts, dynamic and simple. In addition, a fashionable and modest luxury is displayed through the application of various-style materials in the space. The application of lines in the space is simple and modest. However, the selected furniture and decorations are both fashionable and luxurious, making you sense a strong modern atmosphere.

项目名称：绿地海珀旭辉C5户型
设计单位：大观国际空间设计有限公司
设 计 师：连自成
主要材料：木化石，胡桃木，珍珠木，贝壳壁纸

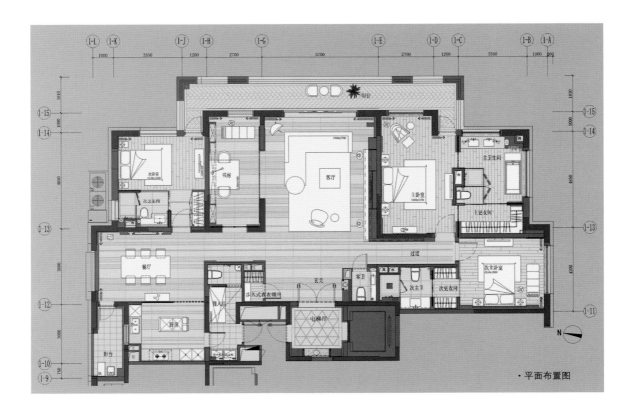

・平面布置图

整体空间以现代低调的手法贯穿，运用分割体块的方式将空间的量体切割成既动感又纯粹的比例，再加上不同表情的材质性格融合在整体空间中，呈现出一种时尚的低调奢华，空间上的线条强调纯粹与内敛，而在家具及饰品的选择上注入一种时尚以及高级的质感，让空间慢慢地渗透出一种浓厚的现代感。

Simple and Oriental Style
简洁东方

The whole space is characterized in modern and concise style, with grey coffee color highlighting the steady main key of the entire space.

Concise and Eastern style furniture creates a certain special atmosphere in the space. Meanwhile, Anglepoise large desk lamp, Him & Her Chair by Casamania and fashionable furniture also become lightspots of the steady space.

The whole space is filled with a warm, comfortable and peaceful atmosphere.

项目名称： 田厦国际中心公寓162平样板间
设计单位： 于强室内设计师事务所
设 计 师： 于强
项目地点： 广东 深圳市
建筑面积： 162 m²
主要材料： 胡桃木、白洞石、墙纸
摄 影 师： 吴永长

顶级样板房 | Top showflats

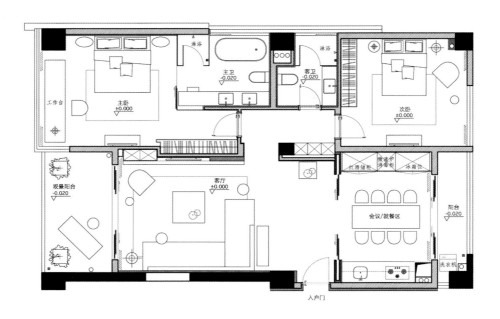

· 平面布置图

设计主要运用现代简洁的手法营造整个空间，灰咖色胡桃木奠定了整个空间沉稳的基调。

简洁东方意味的家私给空间带来些许意蕴，Anglepoise巨型台灯和Casamania的Him & Her Chair现代设计感的家具又给沉稳的空间增加了亮点。

整个空间表达出一种温暖、舒适、宁静的氛围。

Modern City Fashion
都会时尚

The project is situated in the downtown area and the top floor is in duplex style. The duplex room features modernized design style, with prosperity and fashion of modern cities contained in the rooms.

The arrangement of the space is simple, and the parlour background wall is naturally hidden in the two-layer high book cabinet. As for the collocation of furniture, it integrates the modern style with the classic one, the Eastern culture with the Western one, representing the fashionable and modern urban style.

项目名称： 田厦国际中心翡翠明珠复式样板间
设计单位： 于强室内设计师事务所
项目地点： 广东 深圳市
建筑面积： 426.6m²
主要材料： 灰木纹大理石、雅士白大理石、黑檀木 烤钢琴漆、墙纸
摄 影 师： 吴永长

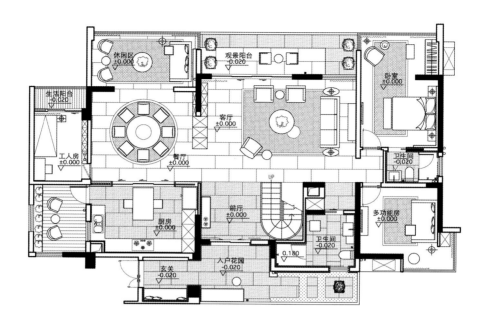

· 一层平面布置图

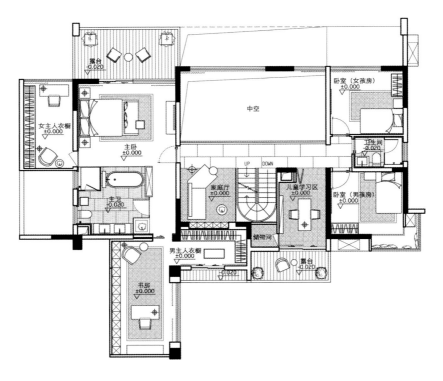

· 二层平面布置图

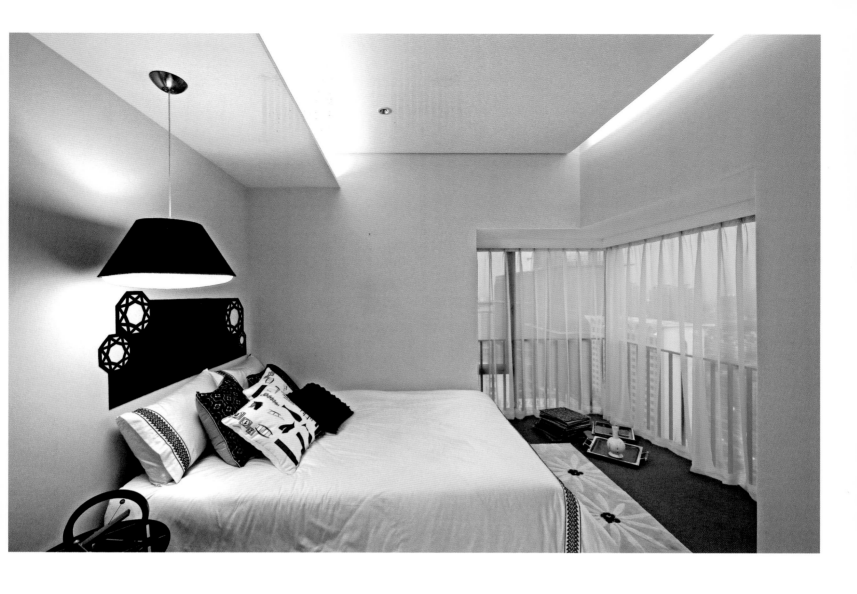

项目位于城市中心区域,为顶层复式。设计主要运用现代的设计理念,将现代都市的繁华与时尚延伸至室内。

空间以干净简洁的建筑体块构造,将客厅背景墙自然地隐藏在二层挑高的书柜中。家具搭配融合了现代与古典、东方与西方的元素,体现出时尚、现代的都会风格。

Northern Europe Style
北欧情怀

Design style of friendships and humanistic of north Europe had been integrated naturally in the flat conception of urban hotel flat. To give feelings of simple, comfortable, natural and fashionable to people by natural realizing living function in limited space. It has reflected quality life of urban young people.

项目名称：田厦国际中心公寓45平样板间
设计单位：于强室内设计师事务所
项目地点：广东深圳市
建筑面积：45m²
主要材料：指纹白、墙纸、苹果木
摄 影 师：吴永长

顶级样板房 | Top showflats

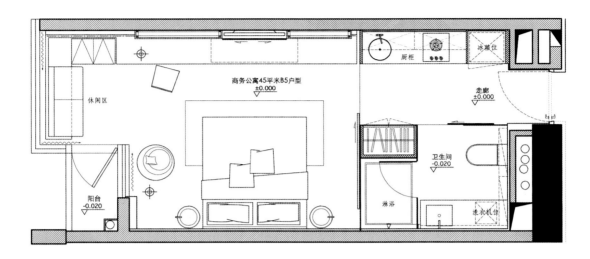

· 平面布置图

将北欧友好、人性的设计风格自然地融入都市酒店公寓的概念中。在有限的空间中自然地实现了生活起居功能,给人简约、舒适、自然、时尚的感觉。

体现了都市年轻人的品质生活。

New Oriental
新东方主义

Fashion is the design theme here. Standing in the room, every time you stop and look up, you can find the balance containing both function and beauty. The living room and dining room face each other. The banyan veneer is used in the bilateral cupboard, completely covering the hidden-style appliance storage.

The carving tawny glass behind the sofa makes the room soft. Classical lines are carved in the television walls, making the room more elegant. Furniture in the dining room is white, showing the elegance and cultural style of the room. The full height grating bookshelf with iron work used for frame, adding mobility to your eye impression. It takes good advantage of the space, letting in enough sunlight, which enables the hosts to enjoy more comfortable rest time when reading.

Apart from the continuous use of wood, the island-style bed frame is introduced into the master bedroom. Neat line is another feature of eastern culture. The secondary bedroom is decorated with concise and modest wooden materials and leaves residents as much space as possible. The whole room is decorated with silver gray geometric totem wallpaper. The introducing of sky blue makes you feel much younger when stepping into the room. The design style makes every space and furniture more flexible, perfectly creating a combined living environment filled with culture, modest and connotation.

项目名称：新东方时尚英桥帝景
设计单位：伏见设计事业有限公司
设 计 师：钟晴
项目地点：中国台湾省桃园县芦竹乡
建筑面积：270.6m²
主要材料：天然大理石、玻璃、黑镜、版岩木皮、铁件(镀钛、镜面铁件)、木质地板、进口壁布、进口壁纸

顶级样板房 | Top showflats

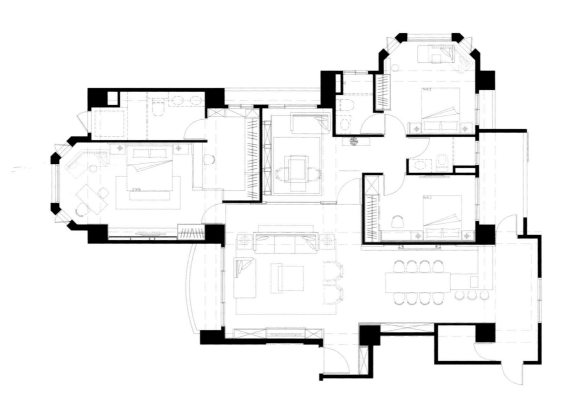

· 平面布置图

以东方时尚为题，立于空间中，每步的驻足、仰望的视角，皆隐含机能与美感的平衡；客、餐厅形成对应空间，应用版岩木皮延伸两侧壁柜，完整包覆住隐藏式收纳柜体。

客厅沙发后方以雕花茶镜营造出柔和空间质感，电视墙面勾勒出古典线条，让空间更显优雅，餐厅家具利用白色色系，更增添出空间的雅致与人文并存的风格；餐厅以黑镜为辅，增加空间感，摆设富含中国元素的画作，特地选用大理石桌与仿明式座椅，并以铝件吊灯为主要灯源，于木质空间含入金属元素；书房则用落地格栅书架，利用铁件造型框架，使视觉印象增添律动性，并利用空间优势，保留大面采光，自然光线让主人在阅读时，能拥有更舒适的休息时光；卧室延续木质色调与手法外，主卧筑起仿海岛床框，利落线条也是东方元素的重要呈现；次卧以简约、低调的木件，并保留空间的最大值；卧室则以银灰色的几何图腾壁纸铺满整室，带入天空蓝的色系，让人进入空间时，感受到轻盈的年轻气息。

此设计中，每个空间、家私及介面扭转了调性，完美打造出兼具修养、内敛、涵养的居住环境。

Modern City
新巴洛克主义

The room is featured by the insight of the combination of classical and modern styles, and set imaginative and fluid art style as its base, and the innovative thinking of the concept of space to provide a rich visual experience for customers, and also add unlimited imagination to the landscape of the ancient city. The metropolitan geographical features find the same exquisiteness with the wild enthusiasm of the Modern Baroque style. The designers eager to create the fluid line art and the visual experience to serve the high-end real estate market atmosphere with a rich sense of luxury; more importantly, it is made for people to observe that today's Chengdu has been on the international stage, and the diverse topography of city connotation transformation to the mixture of Chinese and Western quintessence. Specially, Deco elements were added in the rooms, the use of restraining color balances out the excessive extravagance of modern Baroque. People can find a drizzle of simplicity and plainness in affluence.

The original plane is set to separate the living and dining rooms in which designers link the two rooms on purpose. This breakthrough thinking has made the overall vision even more spacious. Restaurant is designed next to the floor-to-ceiling windows, sp that the living room can be endowed with the complete and stable main wall. When the family gathered to celebrate the happy time while enjoying the multiple landscape outside the window, what a spectacle of natural beauty and human interaction!

项目名称：成都东大街宏誉地产样板房
设计单位：玄武设计
设 计 师：黄书恒
参与设计：胡春惠、张禾蒂、杨惠涵
项目地点：四川 成都
建筑面积：200m²

顶级样板房 | Top showflats

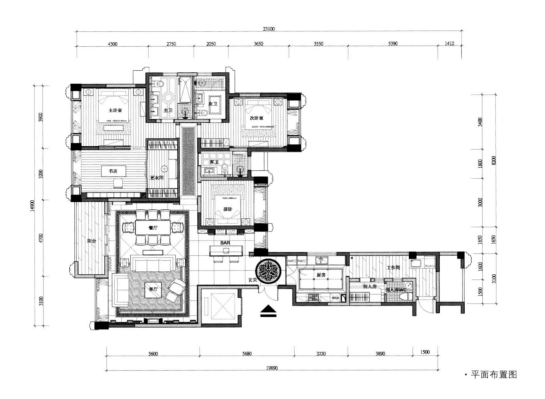

· 平面布置图

本作品体察古典与现代交融之必然，以奔放流畅的艺术风格为基底，同时借由空间概念的创新思考，为客户提供丰富的视觉感受，亦为古老都市的景貌，添加无限想象的可能。繁华的地域特性，与热情狂放的现代巴洛克风格，有异曲同工之妙，设计者渴望借由流畅的艺术线条，与富有奢华感的视觉配置，服膺高端房产之市场氛围；更重要的，是细致地体察到当今成都市已跃上国际舞台，城市内涵转化为中西并陈的多元形貌，本作品特地导入装饰主义之元素，利用内敛的色彩运用，收束现代巴洛克的过度豪奢，在富庶之中，透露几许质朴气息。

原始平面设定为客厅与餐厅分隔，设计者特别打通两个空间，这个突破思维框架的做法，使得整体视觉更加开阔。同时将餐厅设置于落地窗旁，客厅可获得完整稳定的主墙面，当家人齐聚一堂，欢度时光的同时，也享受着窗外的多重山水，天光隐入云影之间，景色与笑语相偕相伴，体现了人文与自然完美交融的细致考虑。利用镜面与金属质感材料，营造出堂皇明亮的感觉，衬托室内的简单色彩，同时增加了视觉的深度与广度。

Culture and Life
文化栖居

To continue architectural style of architecture ARTDECO and bear classical marrow, after having solved function rationality by the design to interior space, how to construct temperament aesthetics of eastern thinking, how to convert the aesthetics to space, constructing internal order consistency both of cultural temperament and functions form.

To conclude living space essence such as both visible and invisible things of sunshine, wave, green plants, free air, joyful and happy, to explore temperament aesthetics of eastern space, emphasizing building both of cultural atmosphere and spirit belongings.

Both on furnishings and accessories, starting from eastern cultural background, using eastern elements both on varying degrees and force (bamboo, chinaware, paintings of Huaiqing Wang, silk fabric etc) to overturn routine view of everybody on original, appear to contrast between material quality itself and background, deliver cultural attribution and still bring eastern emotional experiences to habitants at the same time of owning international aspect.

项目名称：千灯湖一号
设计单位：HSD水平线室内设计
设 计 师：琚宾
参与设计：韦金晶、邱建军、陈建君、刘佳
软装设计：石燕
项目地点：广东佛山
建筑面积：300 m²
摄 影 师：井旭峰

延续建筑ARTDECO的建筑风格，承载古典精髓，室内空间的设计在解决了功能合理性之后，如何去建构东方思想中气质美学，如何将这种美学转化在空间之中，如何协调文化气质与功能形式的建构内在秩序的一致性。

归塑居住空间本质，如阳光、水体、绿植、自由的空气、愉悦、美好等等有形和无形的体。从而探寻的东方空间的气质美学，着重的是文化氛围和精神归属感的营造。

在陈设配饰上，以东方文化背景为出发点，通过不同程度和力度地使用东方元素（竹、瓷器、王怀庆的绘画、丝绸面料等等），从而颠覆大家原有的常规看法，通过材质本身和背景的对比，和文化属性的传递，使其在拥有国际面孔的同时，依然带给居住者东方式情感的体验。

Nature's Charm
自然旖旎

After entering a room, taking sideboards as zoning between a hallway and a message room to highlight decoration effect on hallways and use function. All of living room, dining room and public sections had given priority to wood furniture to work in concert with sideboard. The light design of separate obstructions on sideboards coordinates with elaborately arranging adornments, which are all designed for intensifying depth of field of perspective. We can have a panoramic view on outside garden at living room through pushing wooden shutters, which will make spaces both indoors and outdoors be in complete harmony and in one integrated mass to send out intoxicant natural gorgeous feel.

Stair hall is a bright spot throughout the whole house. Designers applied material quality of Eliza cream-colored marble which is same to the ground at living room, utilized the whole grayish violet scouring pad to make basement integrate with the first floor, used color to well provide for characteristics of Southeast Asia style, meanwhile, carefully placed grounded light on stair step to promote individuation characteristics of spaces.

Abundant plant resources which are brought by tropical and moist climate in Southeast Asia for the design to the master bedroom, the first choice is wood. Both the customer bedroom and children room also take Southeast Asia wind as themes, penetrating various elegant design elements to fully show the feelings of graceful and natural.

项目名称：长沙万科—金域华府三期项目H-B样板间
设计单位：HOT CONCEPTS
设 计 师：周达星
项目地点：湖南 长沙
建筑面积：235 m²

进入室内玄关与按摩房以装饰柜为划分区域，既能凸显玄关的装饰效果，又凸显其使用功能。客餐厅及公共部分都以木质家具为主呼应装饰柜，装饰柜上的错落的隔断中设计的灯带，配合精心编排装饰品，都为了强化景深透视而设计，客厅透过木百叶推门能将室外花园尽收眼底，使室内和室外的空间水乳交融，浑然一体，散发着醉人的自然绚烂情怀。

楼梯间是贯穿全屋亮点，设计师运用与客厅地面相同的伊莉莎米黄大理石材质，墙面利用整体的浅灰紫打布使地下室与一层连成一体，用色彩丰裕了东南亚风格的特点，同时在楼梯踏步边细心地安放了地灯，提升空间个性化的特质。

主人卧室间设计以东南亚热带、潮湿的气候带来的丰富的植物资源——木材为首选。客卧和小孩房亦以东南亚风为主题，渗入多种雅致的设计元素，把旖旎自然的感觉表露无遗。

Colors
彩条复兴

Bold using wallpapers of vertical color bar at Unit s6 of the paradiso in Foshan, matching with handsome floor lamp of submachine gun style, black dermal sofa, which had carried with a bit steady both in fashion and brief. Open type kitchen linking dining room will greatly save space for you..

The master bedroom on the second floor on the second floor utilized mirror face to stretch the whole space, the cantilevered desk art as well as both chairs and desks of simple art added spirituality to the whole space.

项目名称：佛山万科金域蓝湾S6户型
设计单位：广州共生形态工程设计有限公司
设 计 师：彭征
项目地点：广东 佛山

顶级样板房 | Top showflats

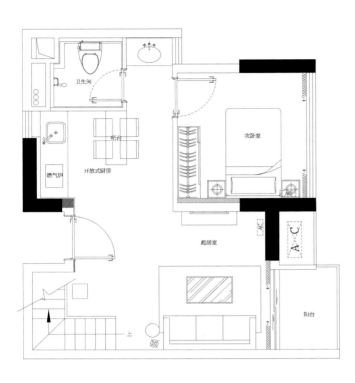

• 一层平面布置图

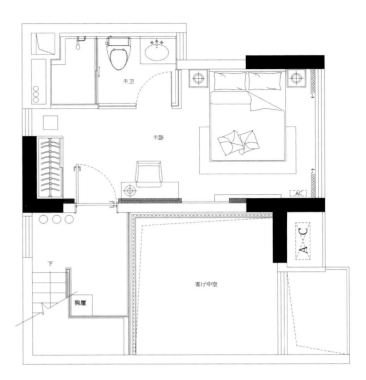

• 二层平面布置图

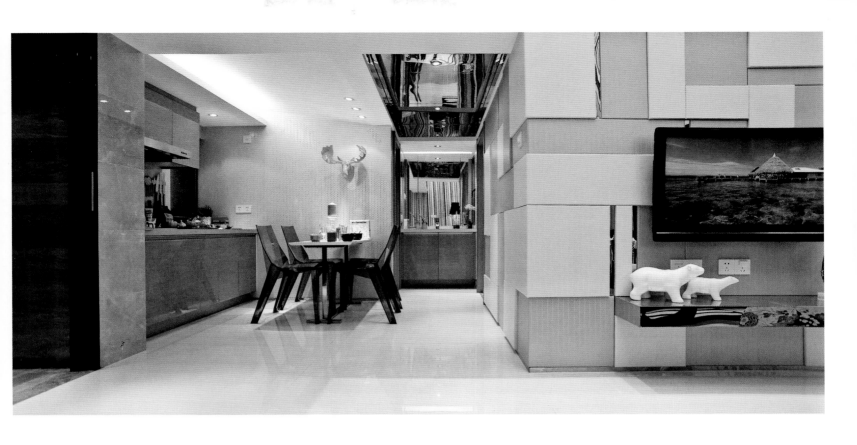

佛山万科金域蓝湾S6户型客厅大胆地采用竖向彩条墙纸，配以帅气的冲锋枪造型落地灯，黑色真皮的沙发，时尚简约中带有几分稳重。开放式的厨房连餐厅设计，大大节约空间。

二楼主卧利用镜面拉伸了整个空间，悬挂式的书桌艺术简约的书椅，为整个空间更添灵动。

Courtyard Garden
天井花园

The courtyard on the first floor had been extended down to another layer, the original closed basement had been added a patio garden, both lighting and venting of whole basement had been improved, because functional orientation to basement is entertainment space, we had transformed the interior into teahouse, bar and table tennis room etc, and formed peacekeeping space of coherence form, the landscape was brought in space such as tea house and table tennis room.

Children room on the second floor has been added a SKYLINE sunshine room, the bright spot of the space is skylight through which we can see sky, while soft decoration has taken plane as themes to give false impression that the master was a little boy. Master room is also located on the second floor equipping with supersize rest room, partition between dry and wet, grounded bath is just by the window.

After stepping onto the third floor, a group of crystal ball droplight is brought in eyes, the level of scattered wave, crystal clear, which has become bright spot in the whole space. A guest room is located on the 3rd floor and a life balcony which is transformed into a roof garden outside.

项目名称：棕榈彩虹花园别墅样板房
设计单位：广州共生形态工程设计有限公司 / COCOPRO.CN
设 计 师：彭征
项目地点：广东中山
建筑面积：415 m²
主要材料：橡木实木地板、薰衣草木饰面、米黄大理石、镜面不锈钢、枯木、棉麻、仿真花

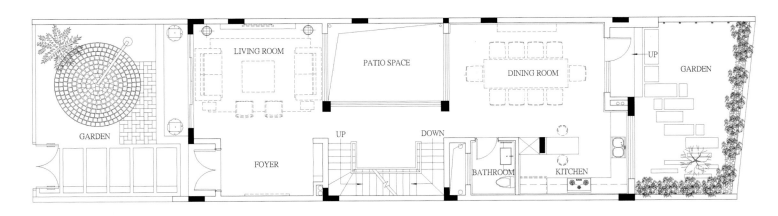

· 一层平面布置图

一楼的中庭往下延伸了一层，原本封闭的负一楼被加建了一个天井花园，整个负一楼的采光和通风得到了改善，由于负一楼的功能定位是娱乐空间，所以我们将室内改造为茶室、酒吧和桌球室等，并形成一个内聚形的维和空间，景观被引入到茶室和桌球室等空间。

一楼改造后的餐厅可以容纳十人的餐桌（原餐厅旁有间客房，餐厅扩大后客房取消），餐厅旁有开放式厨房和超大的红酒柜。客厅和餐厅向外皆连接前后私家花园，向内连接中庭，采光和景观都很好。

二楼儿童房加建了一个SKYLINE的阳光房，看到星空的天窗是空间的亮点，而软装方面以飞机作为主题，假想主人是一个小男孩。主人房也设在二楼，配有超大的卫生间，干湿分区，落地浴缸就在窗边。

走上三楼，映入眼帘的是一组水晶球吊灯，高低错落，晶莹剔透，成为整个空间中的亮点。三楼设客房一间，还有一个户外的生活阳台，被改造成屋顶花园。

Elegant Space
素雅格调

Being steady and restraining, the space had given off rich flavor of simple but elegant, decoration of Chinese style. Appliance and paintings on the wall had brought out the best in each other, which had revealed modern fashionable style and characteristics while retaining traditional implication. Reasonable usage of space, clear zoning of space but correlating each other, which had brought comfortable inhabitancy experiences.

项目名称：大城郎云样板房
设计单位：暄品设计工程顾问公司
设 计 师：朱柏仰
项目地点：中国台湾台中市

空间沉稳内敛，释放出浓厚的素雅气息，中式风格的摆件、器具和墙上的画作相得益彰，保留传统意味的同时，展现出现代时尚的风格特色。空间利用合理，功能区划分明确而又相互关联，带来舒适的居住体验。

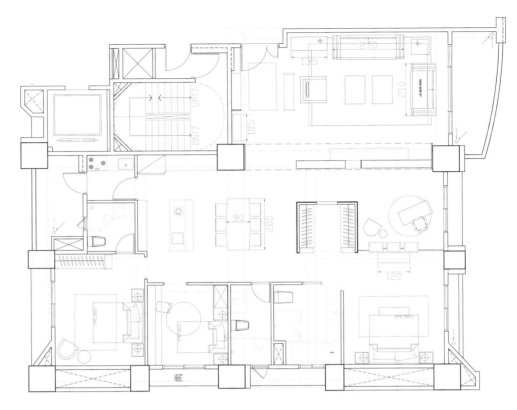

·平面布置图

Zen
禅意东方

Located in holiday resort, Lake Taihu, Suzhou, current project is 180 meters away from Lake Taihu. South of the land masses faces to Lake Taihu, next to culture forum of Lake Taihu, positioned integral figure of the project as both modern and Chinese style.

Closing a gate will namely be deep in a mountain in enjoyable life, Xinjingbian Bodhi has solved problems both of venting and lighting through the concept of "stereoscopic courtyard" to rich layers of building space, making building space own exclusive gardens of experiential style of "stereoscopic courtyard", intensifying residential quality feeling of "eastern courtyard"; meanwhile, pay attention to both interaction and dialogue both of gardens and landscapes between indoor space and outdoor space, the scene will change at every step, which will maximize the limited building resource both of gardens and landscapes; what we had given is a life style.

项目名称：太湖天城别墅样板房之一
设计单位：深圳市昊泽空间设计有限公司
设 计 师：韩松
项目地点：江苏 苏州
建筑面积：500 m²
摄 影 师：江河摄影

本案的整体设计风格为禅意东方。

写意人生"闭门即深山,心静遍菩提"。通过"立体院落"的概念,解决了通风采光问题,丰富建筑空间的层次,使建筑空间拥有"一天一地"的体验式独享园林,强化了"东方院落"住宅的品质感;同时注重室内外空间园林景观的互动与对话,做到移步换景,将有限的建筑园林景观资源做到最大化。我们提供的是一种生活方式。

Water Flows, Wind Breezes
水殿风来

Current project is located at Lake Taihu holiday resort, Suzhou, 180meters away from Lake Taihu. South of the land masses faces to Lake Taihu, next to culture forum of Lake Taihu, positioned integral figure of the project as both modern and Chinese style.

"People say that I live in a city, I doubt that I was on the Land Peach Blossoms", in blatant urbanism, we are eager for heart tranquility for a moment, through clarity of water, wind, being fragrant and moon, prospect both of people and environment, courtyard both of dark green and elegant to demonstrate life attitude both of mature and confident, there is the Land Peach Blossoms in the heart of everybody; a complete set of the project taking water as principal line to run through both the whole space and design, interpreting different "Southeast Asia"

项目名称：太湖天城别墅样板房之二
设计单位：深圳市昊泽空间设计有限公司
设 计 师：韩松
项目地点：江苏 苏州
建筑面积：800 m²
摄 影 师：江河摄影

这次的项目位于苏州太湖度假区,距太湖约180米。地块南侧直面太湖,紧邻太湖文化论坛,项目整体的形象定位为现代中式。本案的主要概念为:水殿风来。

"人道我居城市里,我疑身在桃源中",在喧嚣的都市生活中,我们渴求心灵片刻的宁静,透过水、风、香、月的清澈,人境双绝的意境,青绿翠郁的庭院,彰显出成熟、自信的人生态度,这是每个人心中的桃花源。整套方案以水为主线贯穿整个空间和设计,诠释着不一样的"东南亚"。

Simple Chinese Style
简约中式风

Simple Chinese style is not only the pursuits of excellences and delicacy in detail but also the atmosphere of comfortableness and joyfulness. The design to characteristic house types in the top floor in this case had again highlighted spatial functionality, and the fluency in lines of wall space had presented a turbulent trend, while the selections on hand painted wallpaper had added a bit of grace to the design of rooms. Modern Chinese style is no longer synonymous with antiquity and richness, which was replaced by being intimate, natural, plain, kind and simple, but containing abundant meanings inside.

After entering into indoor hallway, wallpapers and Chinese lattice which are on wall space will immediately highlight the decoration effect of the hallway. Designers used borrowed scenery from landscape design and the permeability of Chinese lattice to enable to see the effect of living room directly, which has an open visual effect and can highlight practical function.

The design of living room and dining room are according to the division of the original building function, in the space of large and uniform, using dark walnut veneer which was added with hand painted ink painting wall paper of Chinese style to express design theme on background wall in living room. Moreover, dark furniture works in concert with colors of walls, and the design of lamps which is like bamboo joint on the ceiling, had strengthened the theme expression.

项目名称：长沙万科—金域华府三期中式样板间
设计单位：HOT CONCEPTS
设 计 师：周达星
项目地点：湖南 长沙
建筑面积：231m²

简约中式是细节上精致卓越的追求,是舒适愉悦的氛围,本案顶楼特色户型的设计再次突出了空间的功能性。墙面线条的流畅演绎出一股汹涌的潮流,而手绘墙纸的选择为房间的设计增加了几分优雅。现代中式风格不再和古老、呆板画上等号,取而代之的是亲近自然、朴实、亲切、简单却内藏丰富意涵。

进入室内玄关,墙面的壁纸及中式花格即能凸显玄关的装饰效果,设计师采用园林设计的借景与中式花格的通透性,可以直接看到客厅效果,有开敞的视觉效果,又凸显了使用的功能。

客厅与餐厅的设计根据原有建筑功能的划分,在一个统一的大空间之中,客厅主要在电视背景墙上,通过深色的胡桃木饰面加手绘的中式水墨墙纸来表达设计主题,深色的家具与墙面的色彩相呼应,顶面竹节灯的细节设计加强了主题表现。

Butterflies in Gray Shadow
灰影蝶舞

Fashils, luxurious silk materials and mental tawny glass. It is simple, grand and delicate.

The t. The modern and concise dining table and chairs look vigorous in the calm space.

The second floor functions as master bedroom, boy's room and guestroom. The wall of the lift hall is covered with large decorative pictures.

项目名称：保利东语花园样板房-灰影蝶舞
设计单位：广州道胜装饰设计有限公司
设 计 师：何永明
项目地点：广东 佛山
项目面积：144m²
主要材料：黑木纹、水云灰、白木纹、鳄鱼纹木饰面板、软包、墙纸、黑蝶贝壳马赛克

顶级样板房 | Top showflats

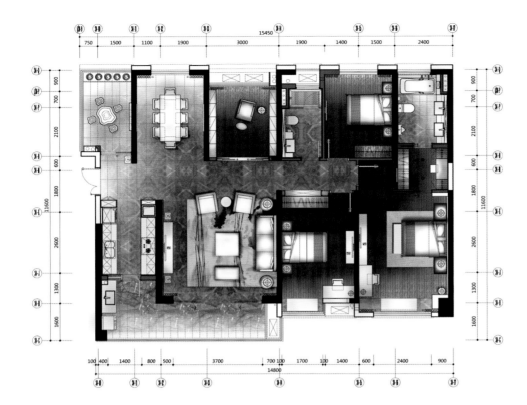

· 平面布置图

整体空间布局规划合理，有良好的采光和通风，开放式的餐厅和客厅一气呵成，既实用又保证了空间的通透性和开阔性，流畅的空间，利落的线条，传达视觉与舒适的平衡。

在设计细节上，都使用"蝴蝶"概念来贯穿主题，各个空间以不同的蝴蝶元素，来体现主题，使空间富有层次变化。在空间的处理上，整体以灰色调为主，木材自然温馨的质感搭配石材温润的色感，相互辉映，使得空间富有节奏韵律感，设计以不矫揉造作的材料营造出低调的豪华感，并塑造出浓厚的文化艺术气息。在空间的艺术装饰品上，线条简洁，颜色彼此对应简洁收敛，注重大小色块间的组合，力求表现自然轻松的情趣，更能符合住宅主人追求独特的诉求。

整体空间设计简约典雅、沉稳中带着新颖，不仅表现了装饰的细腻，还显现出了住宅的温馨与大气，体现主人高雅气质的同时，也传达了在沉稳的基础上也可以悠闲自在的设计观。

American Modern
美式摩登

The space is filled with the modern simple and elegant tonality of American-style with the color of off-white, showing the visual effect of the white board of gallery or showroom and setting off the display of the artist's personal works of art and his collections with convenience to change them at any time. That is the home of a well-known contemporary Chinese American artist.

项目名称：九龙仓时代尊邸30B示范单位
设计单位：香港方黄建筑师事务所Hong Kong Fong Wong Architects & Associates
设 计 师：方峻 TSUN FONG
建筑面积：190m²
项目地点：四川 成都
主要材料：大理石、墙纸、木

顶级样板房 | Top showflats

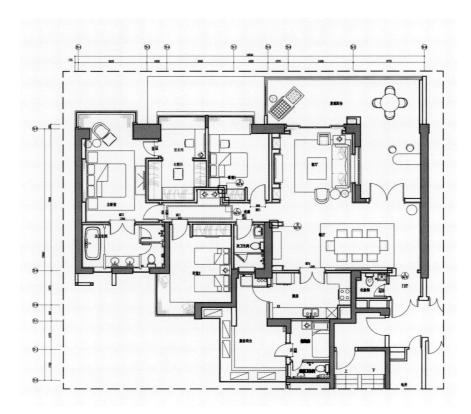

· 平面布置图

充盈着美式摩登的、素净淡雅的米白色系空间调性，让空间拥有如画廊、展厅白色布景板般的视觉效果，映衬出室内陈设的艺术家的个人创作作品或收藏品，并方便随时更换。这便是一位美籍华裔当代知名艺术家的家。

Classic Noble
古典尊崇

We are trying to interpret the essence of Italian neo-classical style, the whole design, in a more welcoming approach, is created to achieve spectacular visual effects, so that people can be more interactive in the space. The neo-classical style has more Italian 18th-century elements to bring out the spirit of the space with more manifestations. Living space is designed to give a respected sense to people, the elements of the black and white stripes shows a combined atmosphere of both modern and classical, the decoration and embellishment of the middle of the space shows an excellent visual effect. The moving lines in the space are designed to allow people to walk in space more casually and naturally, without being interfered too much nce, thus the experience is self-evident.

作品名称：远航.紫兰湖国际高尔夫别墅A3
设计单位：香港方黄建筑师事务所 Hong Kong Fong Wong Architects & Associates
设 计 师：方峻 TSUN FONG
建筑面积：690m²
项目地点：福建 泉州
主要材料：大理石、墙纸

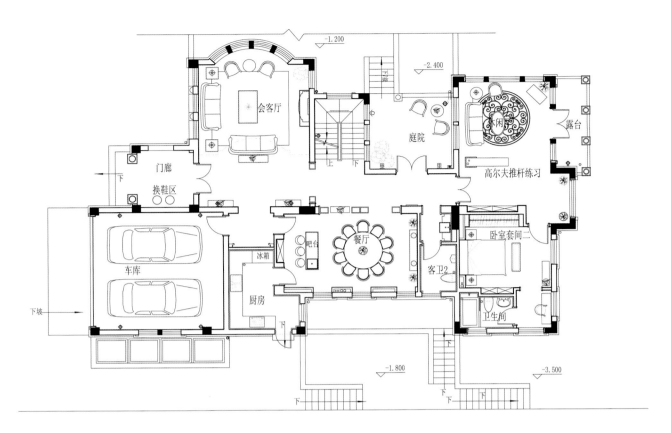

· 一层平面布置图

我们试图诠释出意大利新古典的格调，整个设计以更加温馨的手法来营造出引人入胜的视觉效果，使大家与空间有着更多的互动。新古典风格有着更多的意大利十八世纪元素来衬托出空间的精神，也有着更多的表现形式。起居空间设计给人以尊崇感，黑白相间的条纹的元素呈现出的是一份既现代而又古典的气韵，中间的装饰与点缀使空间呈现出更加良好的视觉效果。而空间的动线设计让人们走在空间中可以更加随意而自然，不至于受到过多的干扰，个中体验不言而喻。

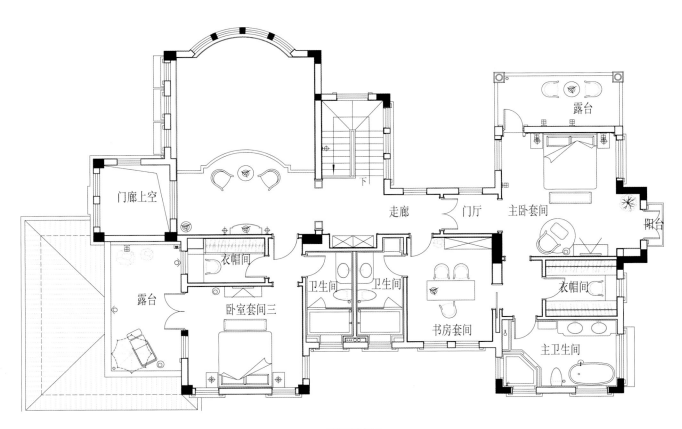

• 二层平面布置图

Low-Key Luxury
低调的奢华

In this case keeping the room's color cleaner and modeling generous. The main of material is woodiness and natural stone which serves as a foil to the style of the theme and do not break it that making space keep regularly. The design reflects the modern Europe style which has less magnificent decoration and powerful color. Giving us a piece of pure, fresh and relaxed space and combining with the characteristics of modern Europe style in each other.

All the wall are adhibited the white oak fillet and wood veneer. Europe style proportional pure and not makes it public that in line with modern life taste. The ground novel stone increases clever element for space that making a primary and secondary fitting supplement with the pure wall. The visual window design used Europe style frame combined with abstract English word as the background wall body, with the modern sofa making one typical European world and taking a European travel mind. In the bedroom the designer broke the sitting room's Europe style, used the dark tonal to match, combined with the downy lamplight and let the bedroom feel quiet and warm. The designer made the low-key costly style in space again and again.

项目名称：顺德均安尚苑
设计单位：广州市柏舍装饰设计有限公司
项目地点：广东顺德
建筑面积：160m²

顶级样板房 | Top showflats

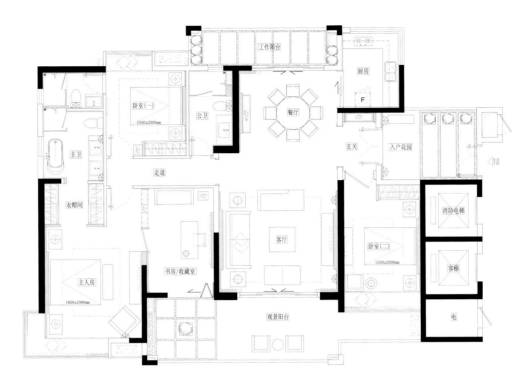

· 一层平面布置图

本案中色调纯净,造型内敛大气,多应用木质元素以及铺贴纹理自然的石材,衬托出风格主题的同时又不失别致,做到了收放有度。体现出的现代欧式风格少了富丽堂皇的装饰和浓烈的色彩,呈现的是一片清新典雅的轻松空间,每一处在设计上都以现代欧式相结合,互相衬托彼此的特点。

整个房子的墙面只铺贴了白色的橡木饰线及木饰面,细致轻巧的欧式比例纯净而不张扬,符合现代生活的品味,地面新颖的石材地花拼贴给整个空间增加了灵动的元素,同时与纯净的墙面壁形成主次合宜的配搭,亮眼处用欧式的画框结合抽象的英文背景墙形体,加上造型简洁大方的沙发构成了一个典型的欧洲世界,送给心灵一次欧洲的游历。在卧室设计中打破了客厅那种明亮的欧式风格,采用了稍微深色的色调来搭配,结合柔和的灯光,让卧室感觉非常安静温馨,是设计师对低调奢华风格的一再延续。

Private Garden
私密花园

For the plane layout, the designer hopes to extend its advantages after analyzing the original building as well as adjusting for the disadvantages of the plane function in order to satisfy the use demands of the buyer from the function. Firstly, entering from the home garden, passing through the porch and presenting the 6m high overhead living room, then the design tries to preserve the clean wall and the French door of the living room that introduces the outdoor lake views. The dining room space is on the first floor. The designer makes an opening between the kitchen and the dining room so that the space has better exposure and shows the style of the big residence. After passing through the half-floor stairs, the mezzanine (second floor) is where the master bedroom is located, connecting the study and cloakroom as well as the independent bathroom for the master suite. The owner can enjoy the entire floor and the privacy of the master bedroom.

The style design is based on modernity and comfort. The teak and solid wood vertical ribs run through the living room, the dining room, the kitchen, the study on the second floor, the corridor on the third floor and the family hall on the fourth floor, linking up the space of the entire floor. The living room features the Silver Wave stone and uses its natural texture as the main decorative wall, reflecting the style with integration of nature into design. It combines the modern oriental interior decorations to enrich the spatial level. The elegant Chinese decorations enhance the style of the show flat.

项目名称：肇庆新世界花园四期
设计单位：广州市柏舍装饰设计有限公司
项目地点：广东 肇庆
建筑面积：298m²

顶级样板房 | Top showflats

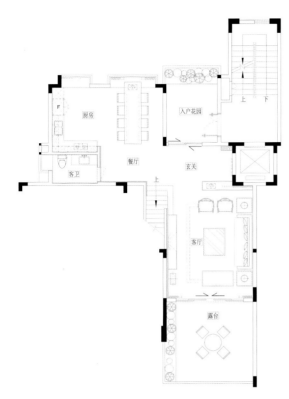

· 一层平面布置图

· 二层平面布置图

在平面布局上，本案设计通过对原建筑的分析，希望把原建筑的优势尽可能扩大，同时把平面功能上的缺点作调整，以达到从功能上满足买家的使用需求。首先，从入户花园进入，通过玄关，6m层高的中空大客厅会出现在眼前，设计尽量保留墙体的干净，以及客厅落地玻璃门的设计，将室外湖景引入室内。首层还有一个餐厅的空间，设计把厨房与餐厅打通，让空间更通透，显出大宅的气派。通过半层楼梯后，夹层（二层）是主人房独享的一层，连通书房和衣帽间，还有独立的主人套房卫生间，使业主可以独享全层之余，也保证了主人房的私密性。

风格设计上，设计以现代舒适为基调。利用柚木实木线条以竖肋贯通中空客厅、餐厅、厨房、二层书房、三层的过道，四层的家庭厅也会出现实木竖肋的设计元素，使全层各个空间得到贯通。客厅主幅以银海浪为主材，利用石材的天然纹理作为主要装饰墙，体现自然与设计融为一体的风格。最后，配合现代东方的软装搭配，丰富了空间的层次，以雅致的中式饰品，提升了样板房的格调。

Clouds
云海之居

Enjoy Peaceful Heart without Material Desire
Skillfully combine the artistic elements and symbols of aesthetics of clan of Hangzhou and devote to professional design completely. The orderly parted space is more prominent, pure, quiet and idle.
"Simplicity" seems having nothing to do with "luxury mansion", however, it reveals their relation to the world farthest by the house with sea view. The graceful embodiment of Zen cannot restrain its nobleness. It is a peaceful and tolerant place without a need to pile up objects.It purifies your body and soul.Why not enjoy the leisure and comfortable life in a place with sea views.
Rain or Shine, Let's Admire the Course of Time
The excellent location with sea in the front and mountain at the back provides elegant atmosphere to private coast. The space has employed perspective skill in large area to interact external landscape with internal court, where you generate different meaning for the things you see. Stay or not, staring clouds outside lazily. It does not only reflect the personal taste on art collection,but also interpret the leisure space and decent style in a classic manner.
Enjoy Freedom of Life with Books and Music. The classic furniture with value for collection and good shape brings topics for the life in luxury mansion and brings delicate taste to life. Following the aesthetic heritage of clan of Hangzhou, the design demonstrates the aesthetics of clan of Hangzhou naturally in every scene as a piece of picturesque.Unrestrained and refined Chinese ink painting, where making villa the exquisite and breathtaking space art.

项目名称： 香水君澜海景别墅
设计单位： 大勺国际空间设计
设 计 师： 陈亨寰、张三巧
软装设计： 上海太舍馆贸易有限公司
项目地点： 海南 三亚
建筑面积： 881m²
客　　户： 海南香水富豪旅业开发有限公司
撰　　文： 刘慧瑛

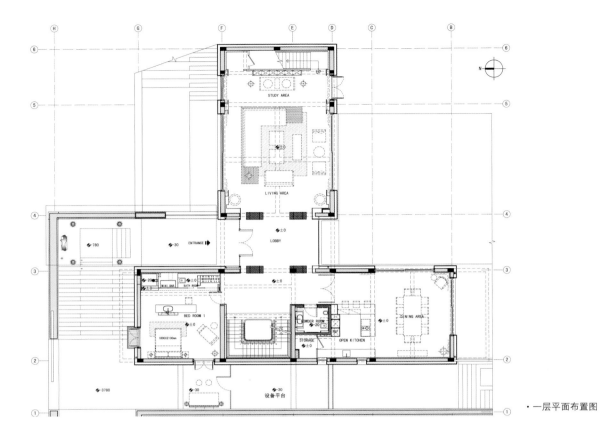

· 一层平面布置图

顶级样板房 | Top showflats

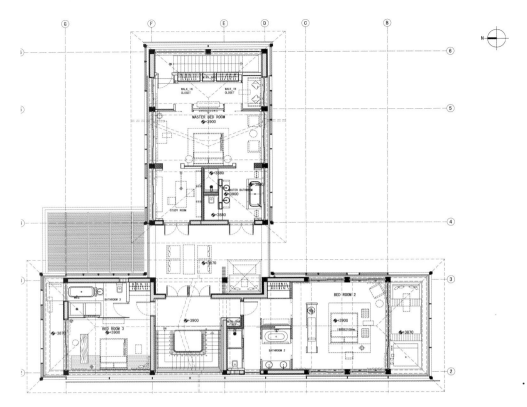

• 二层平面布置图

心无物欲，即是秋空霁海

本案将杭派美学中各式艺术元素与符号加以会贯通，并倾注于团队专业设计力。区隔出井然有序的层次空间，显出更纯粹、更隐逸、更静谧的空间气度。"朴素"与"豪宅"看似漠不相关，却在君澜海景处尽情彰显。宠辱不惊，闲看庭前时间幻化背山面海的极佳地理位置，私属海岸的意境幽雅。空间多处运用大面积透视效果，将外部景观与内庭雅院相互交迭。亦有看山不是山，看水不是水的禅意体悟。

去留无意，漫随天外云倦云舒

继以杭派美学为底蕴，此次设计更是让杭派美学自然而然地展示于各个端景中，宛如描绘一幅行云流水、潇洒脱俗的泼墨中国画，将别墅塑造构建成一座精致而大气的空间艺术品。坐有琴书，便成石室丹丘。具有收藏价值又形神兼具经典的臻品家俬，为豪宅生活的话题性，带来精致的生活情趣。臻品家俬，不仅是体现艺术品收藏的个人品味，更是写意空间休闲风格的经典诠释。

Xi'an Family
熙岸世家

A novelty in the scope, bold creation with excellence, accomplish at ease and be skillful at length. Perhaps this is the holy canon pursed by the designers, which focuses on the use of contrast and contradiction for the achievement of another tranquility. Designers set such rational design practices as standard in the decoration to pave the way for space within the silence. The simple beauty of artistic conception enables the viewer to feel the human temperament without any artificial addition. It also highlights the the beauty of the space itself, thus the organic architectural space is filled with dynamic direction, rather than pursuing the dazzling visual effects. Moreover, it highlights the sense of living of the occupants with the restoration of natural beauty and strength of the building. No matter how it has been innovated, millions of the essence can be converged in a single style without being too much extroverted or introverted. Hecnce he seemingly ever-changing and tense space presents a stabilizing force, exciting yet totally natural. Designers have made the deep space connotations reveal to people with just a few lines of understatement.

项目名称：中海熙岸世家
设计单位：HSD水平线室内设计
设 计 师：琚宾
项目地点：江苏 苏州
建筑面积：337 m²
主要材料：西班牙米黄、金世纪、帝黄金、钨钢、钛金条、木饰面、彩色玻璃
摄 影 师：孙翔宇

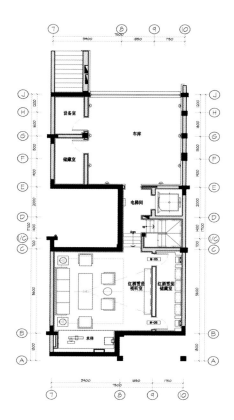

· 负一层平面布置图

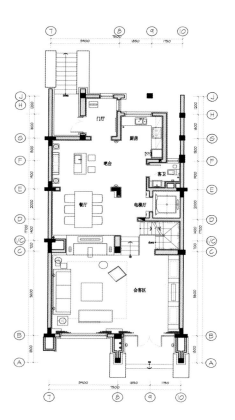

· 一层平面布置图

本案注重运用对比与矛盾所产生的反差,以寻得另一种极致的宁静。设计师用这种理性的设计手法作为装饰的法度,为空间内铺垫出寂静朴素的意境美,让观者在进入空间之初就能感受到这份未经雕琢的人文气质,亦突出了空间自身的机构之美,让有机建筑空间充满着动态的方位诱导,而非追求耀眼的视觉效果。更多彰显居住者的活动本身,还原建筑自然的美和力量。

不论如何创新,都能将万千精华约定在同一法度之中而自成一派,无过无失,让看似多变而富有张力的空间呈现稳定的力量,精彩而浑然天成。设计师于轻描淡写间,洞悉出深邃的空间内涵,禅意无边。

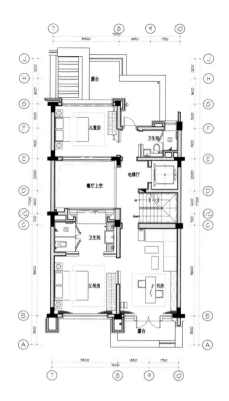

·二层平面布置图

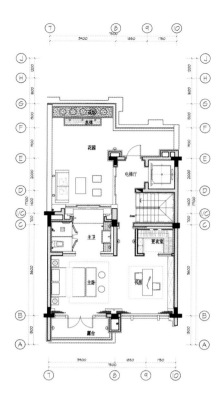

·三层平面布置图

French Style
法式风情

This showroom is the upscale villa premise of the famous brand Rong Qiao and villa of the second phase of Qishan and Wencheng of Rong Qiao. The architecture is French style. There are five floors together counting the top one and the usable area is nearly 1100m².

The simulation owner of the showroom is a family of three generations with 6 people (the couple, parents, a daughter and a son). According to the above family structure, the parents' room, boy's room and girl's room are arranged on the third storey each with a restroom and dressing room. The master's room is designed on the fourth storey and the top storey. There is a master room, a dressing room, a big restroom and a study on the fourth storey. The top storey is designed to be master's family room and collecting room of rare and famous things. There is a guest room, a drawing room, Chinese and western dinning room and family room on the second storey. There is an audiovisual room, red wine and cigar area, relaxation area, tea room, chess and card room. In addition there is a housemaid room and related area. Use the courtyard of the first storey as family barbecue area to enrich life.

In design style, it continues the essence of European style of building exterior fa?ade and it is luxurious and opulent by adopting the technique of combining elegance with modernism. Its' main materials are natural marble, tapestry brick imported from Spanish, antique mirror face, imported wallpaper, imported solid wood floor and customized luxurious carpet and so on. Soft furnishing style mainly uses new classic elegant furniture, lamp, fabrics and diamond glass decoration utensil to create the villa space with low profile and luxury suiting the middle-aged owner's taste, temperament and status.

项目名称：融侨旗山文域二期别墅样板房
设计单位：福州林开新室内设计有限公司
设 计 师：林开新
参与设计：陈青青、陈晓丹
项目地点：福建 福州
建筑面积：1000m²
主要材料：大理石、护墙板、墙纸
摄 影 师：吴永长

本样板房为品牌地产融侨集团旗下的高端别墅楼盘,融侨旗山文城二期别墅,建筑风格为法式建筑。加阁顶层共5个楼层,实用面积近1100m^2。

本样板房的模拟业主为三代六口之家(夫妻两人,父母两人,一个女儿,一个儿子)。依据以上家庭结构,平面功能在三层布置了父母房、男孩房、女孩房,各自带有卫生间、更衣间。主人房布置在四层及阁顶层,四层本设计有主人卧室、更衣间、大卫生间、书房,阁顶层设计为主人家庭室及名贵收藏室。在二层设计有客房,客厅,中、西餐厅,家庭室。一层设计有视听室,红酒雪茄区,游戏娱乐区,茶室,棋牌室,另外设计有保姆房及相关工作区。利用一层庭院设计为家庭烧烤区,丰富生活乐趣。

设计风格上,延续建筑外立面欧式之精华,结合精致,现代混搭的手法来诠释,奢华和贵气。主要用材有天然大理石,西班牙进口饰面砖,仿古镜面,进口墙纸,进口实木地板及定制豪华地毯等。软装风格选用新古典精致家具、灯饰、布艺及水晶玻璃装饰器皿。营造一个低调奢华的适合中年主人品位,气质和身份的别墅空间。

The Post 80S
80后主张

The target population of this design is the people who are born in 1980s. The design idea is also about these people. The different thinking patterns lead to different attitudes towards life. So we break the consistent living mode, position our living with 1.5 persons and create the living space which is suitable for the people who are born in 1980s.

Full opening is unequal to nothing. As for the people who are born in 1980s, the kitchen is not necessarily used for cooking; sofa is not necessarily used for sitting seriously; shower cubicle is not necessarily used for bathing; bed is not necessarily used for sleeping. So, dining room and kitchen are integrated. A long bar counter with cookers can satisfy the requirements of cooking and feeding themselves. The parlour and dining room are in the same space. Sofa is enclosing-type.

Full intelligence is not equal to laziness. Not only alarm clock is annoying grievous news, but also slow rolling of electric curtain, temperature regulation of air conditioner, gradual starting of light, real-time display of weather forecast and financial consultation and so on. We are born in such an age. Don't we betray our "laziness" if we do not use high technologies? Regardless of your criticism, we live more comfortable and convenient than you..

Worshiping foreign things is unequal to flattering the foreign things. For a long time, the generation after 80s always presents the negative impression. The offensive words can be listened everywhere and we live in the discrimination. We break the profane cage through our action. We love our country, we make intellectual enquiries and we are just and free. Now the leading role of this age is us, the generation after 80s. The generation after 80s is unequal to ignorance, treason, selfishness and no-responsibility. "White" is main color in the whole house. We choose white oak for the ground, white stowing varnish for wall space, white real stone coating, white artificial stone, white emulsion paint for ceiling. In the space, we do not need redundant decorations and treatment.

项目名称：世欧王庄1980样板房
设计单位：广州市东仓装饰设计有限公司
设 计 师：梁永钊
参与设计：黎颖欣 魏敏
项目地点：福建 福州
建筑面积：130m²
竣工时间：2012.12

顶级样板房 | Top showflats

・平面布置图

本案设计以20世纪80年代出生的人群为设计目标,设计思想围绕着这个人群展开。思想模式的不同,导致了生活态度的不同,因此我们打破一贯的居住模式,以1.5人居住定位,创作出适合80后的居住空间。

全开放不等于空无一物。在80年代生人的空间里,厨房不一定是用来做饭的,沙发不一定是要正襟危坐的,淋浴间不一定是用来洗澡的,床也不一定是用来睡觉的。为此,餐厅与厨房是一体的,一张带灶台的长吧台足以满足做小吃及喂饱自己的要求。除了有商业机密的书房与给客人整理自己的客卫外,整套居室内都没有"房门"这一概念,全开放的步入式衣帽间,全开放的主卫与全开放的主卧室。主卫与主卧室是在同一空间里的,一个四面玻璃的通透淋浴间搁在床与洗手台之间。在这样的空间里,充斥着80后思想与生活的自由表现。

整套样板房尽可能"白",材质上地面用的是白橡木,墙面用的是白色烤漆板、白色真石漆、白色人造石,天花用的白色乳胶漆,空间上,能减则减,能收则收,没有多余的装饰,也没有多余的处理。